Dr Cleghorn
21.50

D1348436

Dynamics of proteins and nucleic acids

Dynamics of proteins and nucleic acids

J. ANDREW McCAMMON

M. D. Anderson Professor of Chemistry
University of Houston – University Park
Houston, Texas

STEPHEN C. HARVEY

Professor of Biochemistry
University of Alabama at Birmingham
Birmingham, Alabama

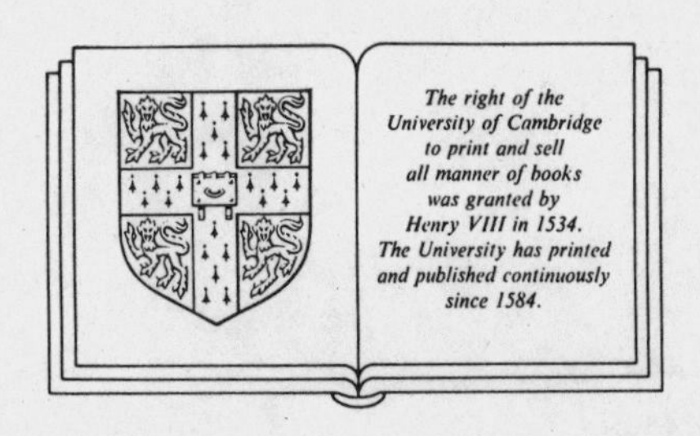

CAMBRIDGE UNIVERSITY PRESS

CAMBRIDGE

NEW YORK NEW ROCHELLE

MELBOURNE SYDNEY

Published by the Press Syndicate of the University of Cambridge
The Pitt Building, Trumpington Street, Cambridge CB2 1RP
32 East 57th Street, New York, NY 10022, USA
10 Stamford Road, Oakleigh, Melbourne 3166, Australia

© Cambridge University Press 1987

First published 1987
First paperback edition 1988

Printed in Great Britain by the University Press, Cambridge

British Library cataloguing in publication data
McCammon, J. Andrew
Dynamics of proteins and nucleic acids.
1. Proteins 2. Molecular dynamics
3. Nucleic acids 4. Molecular dynamics
I. Title II. Harvey, Stephen C.
547.7′5045413 QD431

Library of Congress cataloguing in publication data
McCammon, J. Andrew
Dynamics of proteins and nucleic acids.
Bibliography
Includes index.
1. Proteins. 2. Nucleic acids. 3. Molecular
dynamics. I. Harvey, Stephen C. II. Title.
QD431.M4245 1987 574.19′245 86-17576

ISBN 0 521 30750 3 hard covers
ISBN 0 521 35654 0 paperback

UP

87 08018

Dedicated to
Anne W. McCammon and Marie A. Weaver

CONTENTS

PREFACE

At macroscopic levels, the dynamic character of life is dramatically self-evident. Motion is no less important at the molecular level of biology. Indeed, the marked biochemical effects of temperature changes imply that the activity of biological molecules reflects their thermal mobility. An appreciation of molecular flexibility and dynamics is essential to the understanding of the activity of naturally occurring molecules and to the design of new molecules with specified activities.

Detailed studies of the atomic motion of proteins and nucleic acids are of recent origin. Nevertheless, far more has already been done than can be adequately described in a single volume. The aim of this book is accordingly modest. We attempt to provide the reader with a self-contained introduction to the theoretical aspects of protein and nucleic acid dynamics. The level of presentation is intended to be appropriate for graduate students as well as for research workers in biophysics, physical biochemistry, and molecular biotechnology. Our principal goals are (1) to outline the theoretical methods and their capabilities, (2) to provide a sense of the nature and biological significance of biomolecular dynamics by reference to representative theoretical studies, and (3) to indicate some prospects and directions for future work. Experimental work is covered incidentally in connection with theoretical results.

The book is organized generally to progress from fundamentals to applications and from short time scales to the longer time scales characteristic of most biological activity. Proteins and nucleic acids are treated in an integrated fashion, but mostly in separate sections that can be read selectively if the reader wishes. The first four chapters provide an introduction to conceptual foundations and methodology. Chapters five through eight present the results of selected applications. There, we attempt to describe the nature of the different types of molecular motion that are found to occur in biological systems. The final chapter addresses current research and future prospects. This chapter shows that biomolecular dynamics is entering an exciting new phase, one that is concerned with the interpretation and prediction of biological activity as much as with physical properties. The fruits of this work will include useful tools for pharmacology, medicine and industry.

We hope that this book will give the reader a sense of the special challenges and rewards associated with theoretical studies of protein and nucleic acid dynamics. One of the challenges derives from the fact that these studies involve the fusion of three 'high technology' areas, namely, molecular biology, chemical physics and scientific computing. The rapid pace of development of each of these areas leads to more than the usual rate of obsolescence of research techniques. Such difficulties are, however, more than offset by the promise of aesthetic and pragmatic rewards. Among the former is the pleasure of bridging different scientific disciplines, e.g., using Newton's equations of motion to interpret basic events in biochemistry. The pragmatic rewards include the potential applicability of theoretical methods in the design of new drugs, enzymes, vaccines, etc.

The present book grew out of a review article in *Reports on Progress in Physics* (McCammon, 1984). We are grateful to the Institute of Physics for permission to use material from that article here. Some of the material on nucleic acids is drawn from another recent review article (Harvey, 1986). We also wish to acknowledge our coworkers and colleagues, who have made invaluable contributions to our own understanding of the work described herein. Several colleagues read drafts of the manuscript and made helpful suggestions; particularly valuable comments were provided by Professors P. A. Kollman and B. M. Pettitt, and by Dr T. P. Lybrand, Dr M. Prabhakaran and Dr L. J. Ransom-Wright. Special thanks are due to Denise Marshall for her skillful assistance in word processing. The authors' research in protein and nucleic acid dynamics has been supported in part by the National Science Foundation, the National Institutes of Health, the Robert A. Welch Foundation, and the Texas Advanced Technology Research Program.

Houston and Birmingham J. A. McC.
January 1987 S. C. H.

1

Introduction

1.1 Function of proteins and nucleic acids

Proteins and nucleic acids are particularly prominent among the molecules essential to life. Their importance stems from the remarkable diversity of their functional roles. This diversity can be illustrated by listing a few of the major groups within each of these molecular families. Proteins are molecules that act to build the structural elements of organisms and to provide the energy necessary for life processes. Enzymes are proteins that catalyze biochemical reactions. Familiar examples include the digestive enzymes that degrade foodstuffs to simple, assimilable compounds; the biosynthetic enzymes that build complex molecules from simpler compounds; and muscle proteins that produce mechanical work from chemical reactions. Transport proteins such as hemoglobin facilitate the movement of molecular oxygen and other essential compounds to their sites of utilization. Antibodies are proteins that bind to and neutralize foreign materials that may be harmful to an organism. Other proteins are responsible for maintaining the structures of cells, organs, and organisms, while still others play essential roles in genetic expression, nerve conduction, and all other biological processes. Nucleic acids are the molecules that carry the information necessary for protein synthesis; they can be considered the 'blueprints' that contain the design of the living organism. In both procaryotes and eucaryotes, the genetic information of heredity is carried from one generation to the next in DNA, while various types of RNA's play vital roles in the translation of the DNA sequence of each gene into the amino acid sequence of the corresponding protein. The regulation of the expression of different genes, which is vital to the control of development, growth, repair, and reproduction, involves a wide range of interactions between proteins and nucleic acids.

The functional diversity of proteins and nucleic acids ultimately reflects the large amount of information that is stored in these molecules and expressed in their interactions with other molecules in the course of biological activity. The information content and activity of a protein or nucleic acid can be varied over a large range by modifying its molecular structure. The diversity of naturally occurring molecules represents the cumulative result of structural modifications that have occurred during evolution. The rapid development of molecular biotechnology largely reflects recent discoveries of procedures for preparing systematically modified proteins and nucleic acids in the laboratory.

As suggested by the examples given in the first paragraph, proteins and nucleic acids are largely responsible for the expression and transmission of biological information, respectively, although this division is not a sharp one. As molecular 'machines', an important characteristic of proteins is their specificity of function. A particular enzyme will bind a specific substrate molecule and catalyze a specific chemical transformation of the substrate. A particular antibody molecule will bind specific antigens. For many proteins, the specificity of action is so narrowly defined that a small change in a ligand molecule that binds strongly to the protein (e.g., replacement of a hydrogen atom by a methyl group) leads to a dramatic reduction in binding. The activities of a number of proteins are regulated by interactions with other molecules. For example, their primary activity may be increased or decreased by the binding of specific auxiliary 'effector' ligands. Together with the spatial ordering of proteins imposed by the anatomy of an organism, the specificity and regulability of protein function are largely responsible for the required coherence of biochemical processes. Similarly, nucleic acids have highly specific interactions, both with one another and with other types of molecules, and these interactions are crucial for the control of many aspects of replication, transcription, translation, and recombination. Examples include the specificity of the recognition of the codon on messenger RNA (mRNA) by the anticodon on the cognate transfer RNA (tRNA), site-specific initiation and termination by RNA polymerase, and the remarkable specificity of the restriction endonucleases.

1.2 Structure and dynamics

Given the functional richness of proteins and nucleic acids, one would expect to observe a corresponding complexity in the detailed structure of these molecules. This expectation has been confirmed by X-ray diffraction studies, which have provided the crystal structures of more than 100 proteins and nucleic acids during the past 25 years (Bernstein *et al.*, 1977; Richardson, 1981; Dickerson *et al.*, 1982).

Proteins are very large molecules; their molecular weights are often in the tens of thousands. The basic component of these molecules is the polypeptide chain, an unbranched polymer consisting of a sequence of amino acid residues. There are 20 commonly occurring amino acids, and a typical chain will contain a few hundred of these elementary structural units. Protein molecules consist of one or a small number of such polypeptide chains, complemented in some cases by one or more prosthetic groups (e.g., metal ions or special organic molecules). For a given protein, the polypeptide chain of each molecule is folded compactly into a characteristic three dimensional structure. Although the resulting structures are complicated, it is commonly observed that the packing density of the protein components is nearly maximized, subject to the requirement that those amino acid residues which have a favorable free energy of interaction with water tend to remain near the protein surface. In many cases, it has been possible to carry out X-ray diffraction studies of globular proteins with bound ligands (e.g., substrate analogs). These studies show that the folding of a protein is such that key amino acids with chemically active groups are strategically located in a well-defined 'active site', where the groups can interact in a coordinated fashion with the ligand. Such studies have been invaluable in the development of structural interpretations of protein function.

Nucleic acids are linear polymers whose monomeric units are nucleotides. The size of these molecules covers several orders of magnitude, from tRNA's (with roughly 75 nucleotides, molecular weights around 25000 and end-to-end lengths of less than 10 nm) to eucaryotic DNA's. For the latter, the single DNA molecule of a large chromosome may contain billions of nucleotides and have a molecular weight of more than 10^{12}. If extended, such a molecule would be several centimeters in length. Although tRNA's do have relatively compact structures, large DNA's have extended, wormlike coil configurations at physiological temperature, pH, and ionic strength. Their folding into stable, compact structures *in vivo* is determined by their association with a variety of proteins and, for closed circular DNA, by supercoiling. X-ray diffraction studies on DNA fibers revealed the now classic right-hand double helical structures of A- and B-DNA (Watson & Crick, 1953; Arnott, Smith & Chandrasekaran, 1976). Crystallographic studies have uncovered subtle sequence-dependent variations about these ideal average structures (Fratini, Kopka, Drew & Dickerson, 1982; Shakked *et al.*, 1983), and they have also revealed that alternating purine–pyrimidine sequences can, under some conditions, form left-handed double helices (Wang *et al.*, 1979). A variety of nucleic acid structures other than simple double helices do exist, some transient and others being quite stable. These include single strands, bulged loops, hairpin loops, cruciforms, catenanes, knots, and branches in the double helical structure.

During the past ten years, increasing attention has been focused on the dynamic aspects of protein and nucleic acid structure and function. It has long been inferred from a variety of experimental studies that substantial structural fluctuations occur in these molecules, and that these fluctuations are essential to biological activity (Linderstrom-Lang & Schellman, 1959; Koshland, 1963; Edsall, 1968). Until recently, the exact nature of the structural fluctuations has proved elusive. The recent surge of interest in biomolecular dynamics has largely been stimulated by theoretical studies that have provided a detailed picture of the atomic motion in proteins and, more recently, nucleic acids. These theoretical studies are the primary subject of the present book. The theoretical work involves a combination of methods from theoretical chemical physics and biomolecular structure theory. The methods from chemical physics include techniques that have previously been used successfully to study the structure and atomic motion in dense materials such as liquids and solids. These methods are appropriate in view of the high density and large size of proteins and nucleic acids. Along with the theoretical developments, new experimental techniques that provide detailed insights to biomolecular dynamics have become available. Indeed, the present robustness of this field is largely a result of the interplay of modern theoretical and experimental work. Theory has successfully predicted a number of fundamental properties such as the average magnitude of atomic thermal displacements, the variation of these magnitudes throughout a molecule, and the time scales of certain displacements. Recent experiments have presented new challenges that are stimulating further theoretical work. The results achieved during the past few years and the history of corresponding efforts for systems such as liquids both suggest that the theoretical work on proteins and nucleic acids will become increasingly sophisticated and useful in the coming years.

1.3 Scope of this book

The number of publications on dynamic aspects of biomolecular structure and function is growing at an extraordinary rate. As stated in the preface, this book is not intended to provide an all-inclusive catalogue of this activity. It is, rather, intended to provide a reasonably self-contained introduction to the theoretical foundations of the subject, and to highlight a representative selection of important theoretical results within an integrated framework. The reader may wish to consult recent review articles for additional material (Careri, Fasella & Gratton, 1979;

McCammon & Karplus, 1980*a*, 1983; Karplus & McCammon, 1981*a*, 1983; Levitt, 1982; van Gunsteren & Berendsen, 1982; Cooper, 1984; Edholm *et al.*, 1984; McCammon, 1984; Berg & von Hippel, 1985; Allison, Northrup & McCammon, 1986; Friesner & Levy, 1986; Harvey, 1986; Karplus & McCammon, 1986; Levy, 1986; Levy & Keepers, 1986; McCammon, Northrup & Allison, 1986*b*; Pettitt & Karplus, 1986). A number of experimental results are also described to illustrate the types of data available and the degree of overlap with theoretical findings. Again, excellent reviews that focus on experimental work have recently been published (Gurd & Rothgeb, 1979; Peticolas, 1979; Woodward & Hilton, 1979; Williams, 1980; Jardetzky, 1981; Karplus & McCammon, 1981*a*; Phillips, 1981; Debrunner & Frauenfelder, 1982; Hilinski & Rentzepis, 1983; Huber & Bennett, 1983; Janin & Wodak, 1983; Rigler & Wintermeyer, 1983; Shank, 1983; Wagner, 1983; Bennett & Huber, 1984; Cooper, 1984; Edholm *et al.*, 1984; Englander & Kallenbach, 1984; Middendorf, 1984; Petsko & Ringe, 1984; Torchia, 1984; Friedman, 1985*b*; Ringe & Petsko, 1985; Terner & El-Sayed, 1985).

2

Structure of proteins, nucleic acids, and their solvent surroundings

Before considering the dynamics of proteins and nucleic acids, it is necessary to review some of the structural and energetic properties of these molecules and their solvent surroundings. In the following sections, we sketch some important structural characteristics in simple physical terms. We also describe the interatomic forces that govern molecular structure and flexibility. This general discussion is far from complete. Some additional details are reviewed as necessary in discussing specific dynamic processes in later chapters. Fortunately, many comprehensive reviews of these structural topics are available; references to some of these are given in the following sections.

In considering structure at the level of atomic detail, it is essential to keep in mind the time scale of observation. In macromolecules, there will be subtle differences between any particular instantaneous structure (observed on a time scale that is shorter than the period of vibration of bond lengths) and the average structure that is seen by an X-ray diffraction study which observes average positions over many hours. This is an even more important issue in liquids. An instantaneous structure of liquid water, such as may be obtained from a Monte Carlo or molecular dynamics simulation, will be characterized by well-defined atomic positions. Discussions about this structure in terms of the extent of hydrogen bonding and the similarities to and differences from the structure of ice are possible. Experimentally, however, most methods look at properties averaged over times that are much longer than the characteristic times of rotational and translational diffusion. In this case, only average properties, such as bulk thermodynamic properties, can be determined precisely. In favorable cases, experiment can provide information on the time scales and nature of particular motions, distinguishing, for example, differences in diffusive behavior of water molecules in the bulk solvent and those in hydration layers near macromolecules.

2.1 Water and aqueous solutions

In liquid water, as in other dense molecular systems, an important structural determinant is the size and shape of the individual molecules. When two water molecules approach quite closely, a short range but strongly repulsive force arises from the overlap of their electron clouds. To a good approximation, each water molecule can be thought to have a repulsive, spherical core centered at the oxygen nucleus. These cores do not allow the oxygen nuclei of two water molecules to approach more closely than about 0.24 nm. In simple liquids, such repulsive cores are essentially the only structural determinant. For example, the instantaneous structure of liquid ethane is closely similar to a random dense packing of pairs of overlapping hard spheres, where the overlapping spheres correspond to bonded methyl groups.

Water is not a simple liquid, however. Water molecules have directional attractive interactions that are strong enough to compete with the repulsion of the molecular cores. The most stable molecular configurations still avoid overlapping cores, but have an expanded structure that optimizes the attractions. These attractive forces are of electrostatic origin. Particularly important are hydrogen bonding interactions. When a hydrogen atom is covalently bonded to oxygen, nitrogen, or certain other electronegative atoms, the bonding electron density is partly shifted onto the heavier atom. The hydrogen is left with a significant partial positive charge, and can approach other atoms relatively closely because of the drawn-in character of its electron cloud. Such a hydrogen therefore has a relatively strong electrostatic attraction for oxygen or nitrogen atoms with partial negative charges. These attractive interactions are termed hydrogen bonds; the molecule or group with the hydrogen is referred to as a hydrogen bond donor and the partner molecule or group is referred to as a hydrogen bond acceptor. Hydrogen bonding interactions play an important role in shaping the structure of aqueous systems because these interactions operate only within narrow geometric ranges. The donor and acceptor must be quite close, and the hydrogen can not be too far off the line between the electronegative atoms that it links. For a pair of water molecules with a linear hydrogen bond, the maximum stabilization is about 20 kJ/mol when the oxygens are separated by 0.28 nm; the interaction approaches zero for oxygen separations greater than 0.4 nm.

The electronic configuration of the oxygen atom in H_2O has sp^3

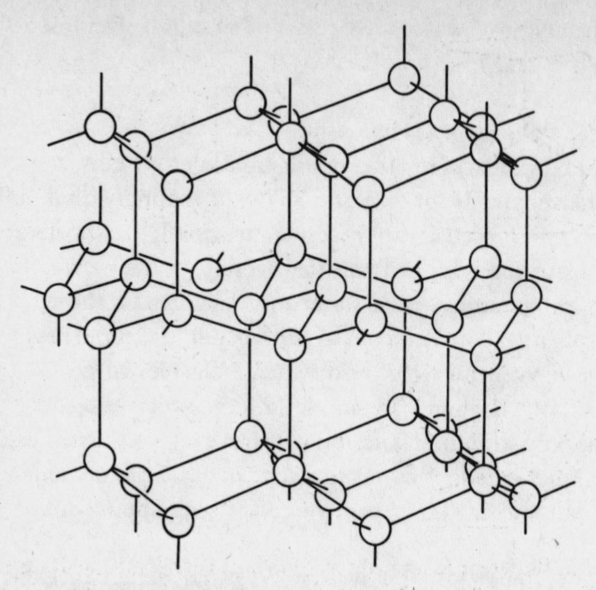

Fig. 2.1. Schematic illustration of the structure of normal ice, I_h. The circles represent oxygen atoms. The solid lines represent hydrogen bonds. In liquid water, thermal energy disrupts the regular lattice structure shown here.

hybridization, with a near tetrahedral geometry that has hydrogen atoms at two of the vertices of the tetrahedron and lone-pair electrons at the other two vertices. Consequently, a water molecule can simultaneously act as a hydrogen bond donor to two other water molecules and an acceptor from two more water molecules. This particularly stable arrangement, in which the central molecule has an approximately tetrahedral set of hydrogen bonds, is replicated in the three dimensional structure of ordinary ice, ice I_h (figure 2.1). In liquid water at 300 K, the average translational and rotational kinetic energy of a water molecule is about 7 kJ/mol, or only about one-third the energy required to break a hydrogen bond. Thus, liquid water retains significant vestiges of the ice I_h structure. Tetrahedral coordination is prominent on a local scale, but broken and defective hydrogen bonds occur frequently enough to destroy the large scale order and rigidity of the crystalline state.

The principal structural features of aqueous solutions of nonpolar molecules can be understood by reference to the local structure of water. When small nonpolar molecules are dissolved in water, the solvent will distribute itself around the solute so as to minimize the breaking of hydrogen bonds. For solutes such as methane, the water molecules in the first solvation shell retain their tetrahedral hydrogen bonding pattern by

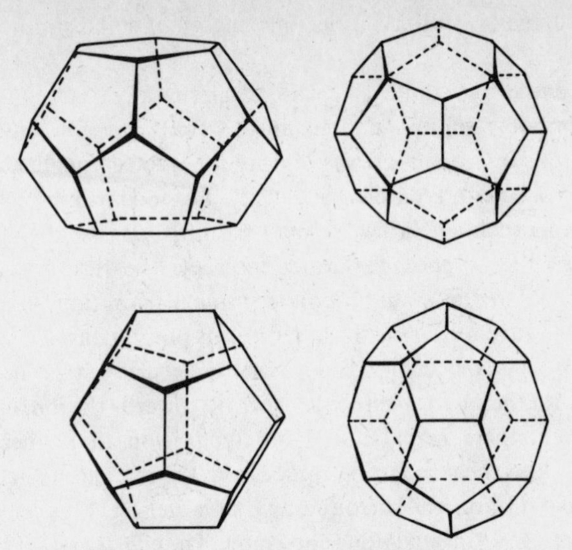

Fig. 2.2. Water structures observed in some crystalline hydrates (Jeffrey, 1984). Oxygens are at the vertices, and the solid and dashed lines represent hydrogen bonds. The hydration cages contain small molecules that are not shown here. The cages that form around nonpolar solutes in liquid water can be thought of as thermally distorted analogs of polyhedra such as these.

forming a cage-like structure around the solute (figure 2.2). Thus, the enthalpic consequences of dissolving a small nonpolar solute are small. There is, however, an unfavorable change in entropy due to orientational restrictions on the solvent molecules. In bulk water, the volume otherwise occupied by the solute would be filled with one or more water molecules. In solution, orientations of the cage waters that would lead to hydrogen bonding with these missing water molecules are energetically forbidden when the solute can not engage in hydrogen bonding. In forming the solvation shell described above, each water molecule straddles a convex portion of the solute surface to form hydrogen bonds with three other water molecules in the shell (figure 2.2). For large solutes, such straddling is not possible for first shell waters that are in contact with flat or concave parts of the solute surface (Lee, McCammon & Rossky, 1984). These waters may therefore suffer some loss of hydrogen bonding. The entropic and possible enthalpic factors outlined above cause nonpolar molecules to tend to aggregate in water, thereby reducing their contact area with the bulk solvent. This apparent attraction between nonpolar molecules in water is called the hydrophobic effect. The hydration effects are quite strong; at 300 K, removal of a single methyl group from water to a nonpolar environment results in a net free energy of stabilization of the order of 10 kJ/mol.

At the other end of the solute spectrum are electrically charged solutes. Whereas hydrophobic solutes tend to move from water into nonpolar environments, ions and polar species display the opposite tendency to a degree that increases with their charge density. For ions with high charge densities (i.e., ions such as Na^+ with small radii, or ions with charges of magnitude greater than one), the water molecules nearest the ion are electrostatically ordered in a tightly bound primary hydration shell. First shell waters of positive ions have both hydrogens pointing away from the ion. If the ion is negatively charged, the water molecules in this shell tend to be oriented with their hydrogen atoms pointed towards the ion. Outside the primary shell, there may be a disordered region that reflects the competition between the structuring forces from the ion and its first shell on the one hand, and from the surrounding bulk water on the other. There is a net attraction between small ions and water. The effective stabilization of such an ion in water is very large, because water molecules have large dipole moments and tend, due to their hydrogen bonding interactions, to reorient collectively in the electric field of the ion. Transfer of a small, singly charged ion from a nonpolar environment to water results in a net free energy decrease of more than 100 kJ/mol; most of this stabilization is due to the favorable enthalpy of forming the primary hydration shell. The properties of ions with lower charge densities differ from what has been described above. Moderately large univalent ions such as the larger alkali ions do not have well-ordered primary hydration shells; instead, the disordered region tends to extend inward to the surface of the ion. Very large polyatomic ions with small charges often exhibit hydrophobic hydration.

The interaction of pairs of ions in water can be qualitatively understood in terms of the phenomena responsible for single ion hydration. The effective electrostatic interaction between a pair of ions at a given separation in water is much smaller than what it would be in a nonpolar solvent. The collective reorientation of water molecules in the vicinity of an ion effectively dissipates the field of the ion, screening its interaction with other ions. For ions that are separated by more than about 1 nm, the effective pair interaction scales inversely with the dielectric constant of the solvent. Typical dielectric constants are 78.5 for water and about 2–3 for liquid hydrocarbons at 298 K; the corresponding electrostatic interactions for singly charged ions of opposite sign at 1 nm separation are -1.8 kJ/mol and -70 kJ/mol, respectively. For ion pairs at separations less than about 1 nm, the effective interaction energy will be modified somewhat by the overlap and mutual perturbation of the ionic solvation shells. The effective interaction of a given pair of ions will be further weakened if salt is added to the solution. Salt ions tend to cluster around any given ion of opposite sign, reducing the net field in the vicinity of the given ion.

It is worth noting here that the effective energies of all of the above solvent-influenced interactions between solutes (i.e., the hydrophobic and solvent-screened interactions) are in fact free energies, since they implicitly reflect a thermal averaging over the possible positions and orientations of the rest of the molecules in the solution. These effective interaction energies are termed potentials of mean force (McQuarrie, 1976; Friedman, 1985*a*, *b*). Like other free energies, these quantities depend on temperature and have enthalpic and entropic components.

More detailed discussions of the structure of water and aqueous solutions of small molecules have been given by Eisenberg & Kauzmann (1969); Bockris & Reddy (1970); Ben-Naim (1974, 1980); Rossky & Karplus (1979); Stillinger (1980); Conway (1981); Friedman (1981, 1986); Geiger (1981); Paterson, Nemethy & Scheraga (1981); Beveridge *et al.* (1983); Jorgensen (1983); Impey, Madden & McDonald (1983); Wolfenden (1983); Marchese & Beveridge (1984); Alagona, Ghio & Kollman (1985); Bounds (1985); Jorgensen, Gao & Ravimohan (1985); Heinzinger (1985); and Rossky (1985). A good introductory survey of this area is provided in the short book by Franks (1983).

2.2 Protein structure

The structure of a short section of polypeptide chain is illustrated in figure 2.3. The polypeptide chain is intrinsically flexible because many of the covalent bonds that occur in its backbone and sidechains are rotationally permissive. The primary limitations to the ranges of rotation for the backbone dihedral angles ϕ and ψ arise from nonbonded interactions. That is, some values of ϕ and ψ are energetically unfavorable because they lead to overlap of the repulsive cores of different atoms ('steric' repulsion). For a typical single residue with a few atoms added to complete the peptide groups at each end, the sterically favorable regions are as shown in figure 2.4.

The residues of which the polypeptide is composed are chosen from the 20 commonly occurring amino acids. A given protein is characterized by a definite sequence of residues; this is termed the primary structure of the chain. The amino acid residues are distinguished by the structures and chemical properties of their sidechains. The sidechains can be divided into two broad classes. Sidechains that are relatively soluble in water are termed

(a)

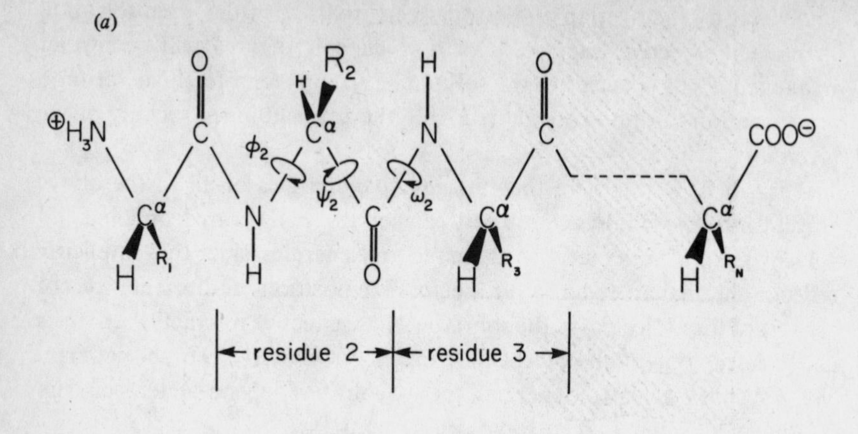

(b)

Fig. 2.3. Structure of the polypeptide chain. The backbone or main chain is shown in (*a*). The covalent bonds and bond angles are rather rigid, but sizable rotations can occur *around* certain bonds. The dihedral angles ϕ_i and ψ_i measure the torsion about the rotationally permissive bonds in the backbone of residue *i*. The dihedral angles ω_i exhibit little variation because the C–N bond has partial double-bond character; each peptide group (CONH) and its adjoining C^α atoms therefore tend to remain in a common plane. The labels R_i represent the sidechains, one of which (a tyrosine sidechain) is shown in detail in (*b*). The tyrosine sidechain has two rotationally permissive bonds; the corresponding dihedral angles are χ^1 and χ^2. The ring remains relatively flat due to partial double bond character in its C–C bonds.

hydrophilic, while those that are less soluble in water are termed hydrophobic. The hydrophilic sidechains include electrically charged groups (acidic groups or basic groups, which typically bear a full negative or positive charge, respectively) and neutral groups with a substantial electrical dipole moment. The hydrophobic sidechains are neutral and relatively nonpolar.

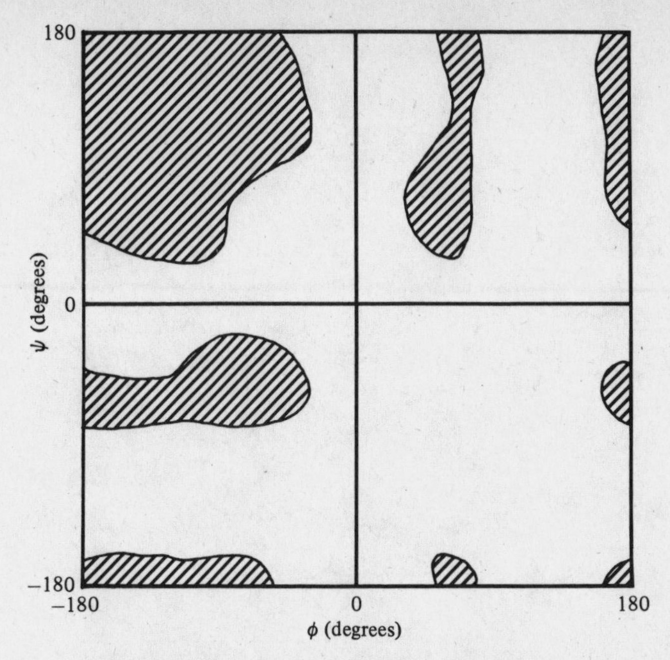

Fig. 2.4. Low energy conformations of an alanine residue in a short peptide chain are indicated by shading. The extended geometry shown in figure 2.3 corresponds to $\phi = \psi = 180°$. The shaded regions have energies within 20 kJ/mol of the minimum energy, based on a model developed by Brant, Miller & Flory (1967). The higher energy regions reflect steric repulsion between nonbonded atoms as originally shown by Ramachandran, Ramakrishnan & Sasisekharan (1963).

As has been mentioned, the polypeptide chain of a given type of protein folds into a characteristic 'native' three dimensional structure in aqueous solution. For simple proteins, folding is a spontaneous process (Anfinson, 1973; Creighton, 1985). The atoms are packed quite densely in typical globular proteins, so that the structures of these molecules can be described in terms similar to those used for dense liquids. Each group of covalently bonded atoms has a size and shape that can be described as a cluster of fused, spherical cores (figure 2.5). The native structure then represents an efficient packing of these groups driven by their effective attractions for each other in the solvent surroundings. In the absence of substantial distortion of this dense structure, the sterically allowed range of local residue motions is much smaller than is implied by figure 2.4.

The attractive interactions that help to determine protein structure include protein–solvent and solvent–solvent as well as protein–protein interactions. The nature of the major interactions involving solvent has already been described. Hydrophobic effects cause nonpolar sidechains to

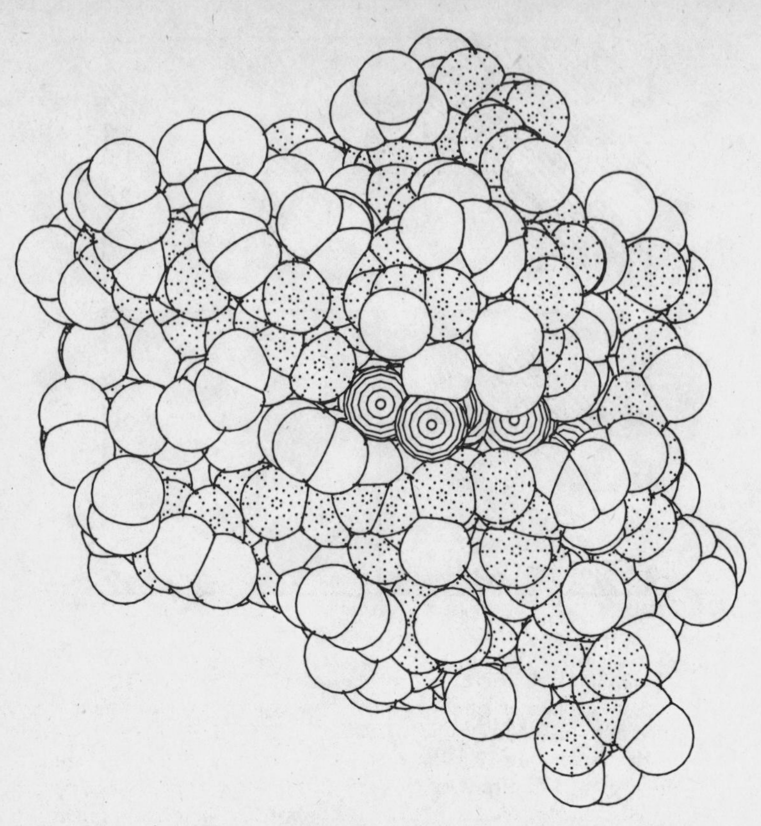

Fig. 2.5. Space-filling representation of the native structure of a relatively small protein, cytochrome c. The sizes of the atomic spheres correspond to the distances of closest approach allowed by the core repulsion forces between nonbonded atoms. (Photograph courtesy of Mike Carson.)

tend to cluster together in the protein interior. This effect is important because many of the residues in a typical protein are hydrophobic. On the other hand, charged groups in the protein will tend to occupy positions at the protein surface where they can enjoy maximal hydration. The geometric features of protein hydration in a highly resolved crystal structure of the protein crambin have been described by Teeter (1984).

The most prominent of the protein–protein attractive interactions is hydrogen bonding (Baker & Hubbard, 1984). Many groups in the polypeptide chain are potential donors or acceptors of hydrogen bonds. These include the peptide (CONH) groups in the backbone, and the sidechains of most of the hydrophilic residues. In the native structure, these groups must interact with one another in such a way that the loss of hydrogen bonds to solvent water molecules is compensated by the

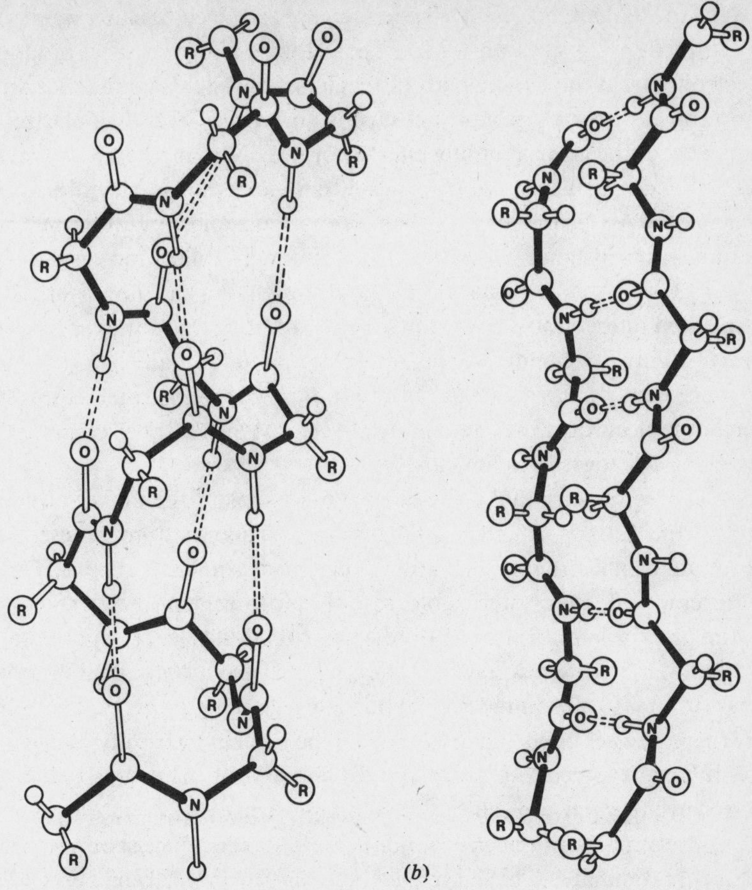

Fig. 2.6. Two types of secondary structure that are often observed in proteins. Both structures compensate for the loss of peptide–water hydrogen bonds upon folding by formation of peptide–peptide hydrogen bonds. (*a*) α helix. (*b*) Antiparallel β sheet.

formation of internal hydrogen bonds in the folding of the protein. For the protein backbone, such arrangements often result in the formation of what are termed elements of secondary structure. A familiar example is the alpha helix, in which the peptide oxygen of residue i forms a hydrogen bond with the peptide nitrogen of residue $i+4$ (figure 2.6(*a*)). Another example is the beta sheet structure, in which extended strands of the polypeptide chain lie next to one another and are cross-linked by hydrogen bonds between their peptide groups (figure 2.6(*b*)). Occasionally, one or a few water molecules occupy well-defined locations within proteins and serve as bridges between charged groups or hydrogen bond donors and acceptors in the polypeptide.

Groups that are not hydrogen bonded are attracted together by weaker, less directional dispersion forces. These forces arise from the motion of electrons about the atomic nuclei in each group. The instantaneous dipole moment of one such group of electrons and nuclei tends to polarize and thus attract the corresponding charges of nearby groups. For two methyl groups *in vacuo*, dispersion forces lead to a maximum stabilization on the order of 0.6 kJ/mol at a carbon–carbon distance of about 0.4 nm; reducing this distance by 0.1 nm results in a destabilization on the order of 10 kJ/mol due to core overlap. Although the maximum attractive dispersion interaction is one or two orders of magnitude weaker than that due to hydrogen bonding, the former occurs among all the groups in a protein. Partly to compensate for the loss of dispersion interactions with solvent molecules in the folding process, the atoms in the protein interior tend to pack together quite closely.

The three dimensional arrangement of groups in the native protein is termed the tertiary structure of the molecule. This structure represents the optimum balance among the various interactions described above. Typical features include the clustering of hydrophobic sidechains in several regions of the protein interior. Charged sidechains tend to remain exposed to the solvent at the protein surface; the few charged groups observed in protein interiors are usually paired with oppositely charged groups. Sections of the polypeptide backbone that are buried in the protein interior typically form secondary structure to compensate for the loss of their hydrogen bonds with solvent water molecules. The packing density of the atoms is quite high, particularly in the hydrophobic clusters where nonspecific interactions predominate. The high packing density is clear in space-filling representations of protein structure (figure 2.5).

Additional discussion of protein structure can be found in texts or reviews by Richards (1977); Schulz & Schirmer (1979); Cantor & Schimmel (1980); Richardson (1981); Creighton (1983); Chothia (1984); Parry & Baker (1984); Perutz (1984); Fersht (1985) and Hol (1985).

2.3 Nucleic acid structure

The composition of a nucleic acid is specified by its primary structure, which is the sequence of nucleotides along the polynucleotide chain. In the case of RNA, there are four constituent ribonucleotides, two whose bases are purines (adenosine and guanosine) and two whose bases are pyrimidines (cytidine and uridine). In the case of DNA, the monomeric units are deoxyribonucleotides. The bases are identical, except that the uracil of uridine is replaced by thymine in deoxythymidine. The structure of a short segment of a hypothetical polynucleotide is shown in figure 2.7.

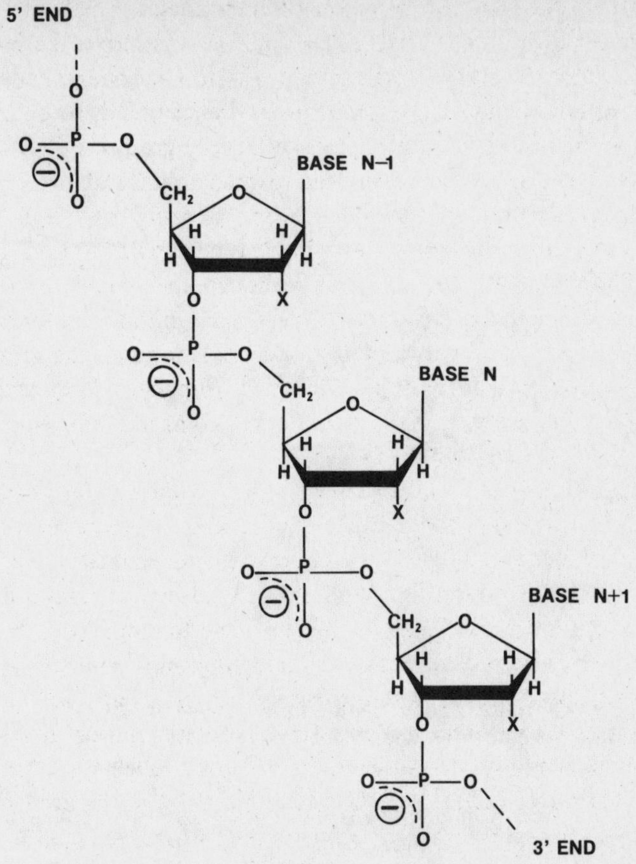

5' END

BASE N–1

BASE N

BASE N+1

3' END

Fig. 2.7. Fragment of hypothetical polynucleotide. Each nucleotide consists of a phosphate group, a ribose, and a base (not shown). Nucleotides are linked by 3′, 5′-phosphodiester bonds. The sequence is numbered from the 5′ end of the molecule toward the 3′ end. For ribonucleic acid (RNA), $X = \mathrm{OH}$, while for deoxyribonucleic acid (DNA), $X = \mathrm{H}$.

The backbone of the molecule consists of alternating phosphate and furanose (ribose in the case of RNA; deoxyribose in the case of DNA) groups, and the bases are attached through the glycosyl links to the C1′ atoms of the furanose rings. The intrinsic flexibility of the furanose-phosphate backbone is suggested by the presence of six single bonds per nucleotide. Torsional rotations about five of these bonds are relatively unrestricted; the C3′—C4′ bond is subject to restricted torsional rotations because of the cyclic furanose ring. The standard IUPAC nomenclature for these backbone torsions is given in figure 2.8. Also shown in that

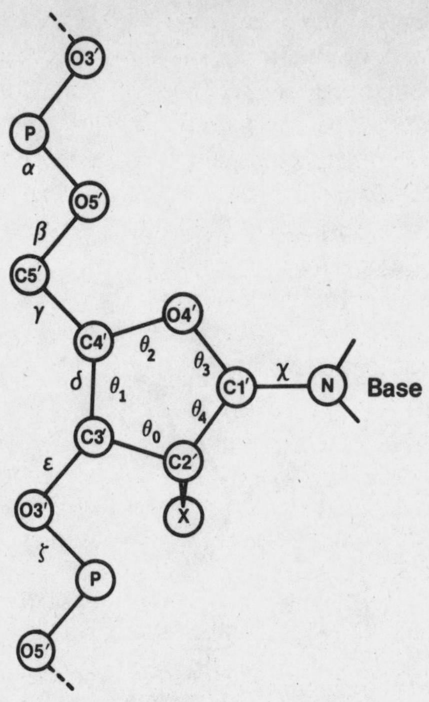

Fig. 2.8. IUPAC nomenclature for backbone torsion angles, ribose endocyclic torsions, and the glycosidic torsion. The backbone torsion δ is defined by the four atoms O3′-C3′-C4′-C5′, while the endocyclic ribose torsion about the same bond, θ_1, is defined by C2′-C3′-C4′-O4′. The proper definition of χ is O4′-C1′-N1-C2 for pyrimidines and O4′-C1′-N9-C4 for purines.

diagram is the glycosyl torsion angle, χ, which determines the orientation of the base relative to the furanose.

In spite of the fact that nucleic acids have more adjustable backbone torsional angles per residue than do proteins (six per nucleotide *vs.* three per amino acid), nucleic acids do adopt recognizable three dimensional structures, just as proteins do. This is partly a consequence of extensive secondary structure within these molecules, which results from hydrogen bonding between bases. This bonding usually follows the scheme proposed by Watson & Crick (1953), whereby guanine binds to cytosine, and adenine binds to thymine (in DNA) or to uracil (in RNA). Alternative hydrogen bonding patterns were observed by Hoogstein (1963) and still others were proposed by Crick (1966). Although Watson–Crick basepairing is the most common, other pairing arrangements have often been proposed in model structures and have been observed crystallographically in both tRNA (Jack, Ladner & Klug, 1976) and in DNA (Kennard, 1985). A second important factor stabilizing the three dimensional structures of nucleic acids is base stacking, which results from dispersion forces and hydrophobic effects. A variety of electrostatic effects also contribute to the stability of nucleic acid structure. These include the site-specific binding of polyvalent cations, particularly magnesium and various polyamines. These cations (sometimes with associated water molecules) tend to bridge between nearby negatively charged groups, especially phosphates. There are also important effects from monovalent cations; these generally do not bind for long periods of time to specific sites but manifest their presence through 'counterion condensation', whereby the statistically averaged properties of the mobile cloud of monovalent cations tends to neutralize the net electrostatic charge of the nucleic acid that arises from the full ionization of one phosphate per residue at neutral pH (Record, Anderson & Lohman, 1978; Manning, 1979; Record *et al.*, 1986). The dynamic nature of counterion condensation results in partial occupancy of specification sites at regions of high electrostatic potential, and the degree of occupancy depends on ionic strength and on the conformation of the nucleic acid. The interplay of these effects is one of the factors contributing to the dependence of conformation on ionic strength (Fixman, 1982; Matthew & Richards, 1984). A final factor that stabilizes the three dimensional structure of nucleic acids is site-specific hydration. This is particularly evident in B-DNA, where a spine of water molecules in the minor groove is observed crystallographically (Dickerson, 1983; Kopka, Fratini, Drew & Dickerson, 1983). The dependence of hydration structure on water activity has been postulated to be another factor in the variation of DNA structure with changes in water activity (Calladine & Drew, 1984).

DNA's are double stranded and double helical molecules, with each base Watson–Crick hydrogen bonded to a complementary base on the opposite strand. Single stranded RNA's also can form helical structures, by folding the molecule in such a way as to bring together short segments of complementary sequences. Three classes of DNA double helices have been observed crystallographically. These include the right-handed A-DNA (Shakked *et al.*, 1983) and B-DNA (Fratini *et al.*, 1982) and the left-handed Z-DNA (Wang *et al.*, 1979). Double helical DNA structures with mismatched G–T and G–A wobble basepairs have also been observed recently (Kennard, 1985), a direct indication of how the flexible double helix can accommodate local structural deviations. Detailed comparative reviews of these structures are available elsewhere (Dickerson *et al.*, 1982; Saenger, 1984). A number of crystal structures have also been reported for tRNA's, beginning with $tRNA^{Phe}$ (Quigley *et al.*, 1975; Ladner *et al.*,

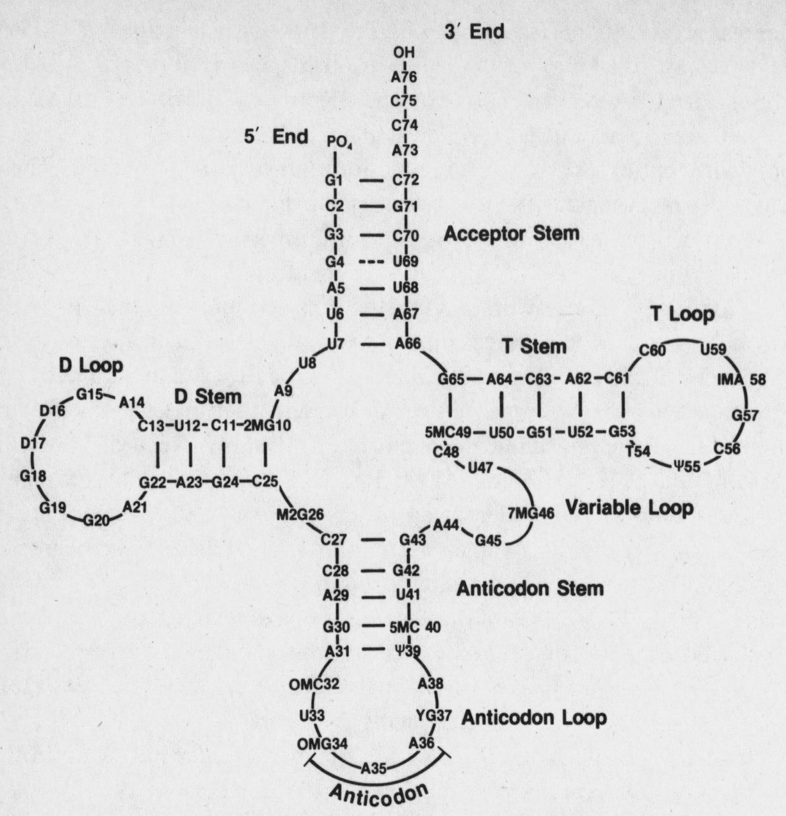

Fig. 2.9. Cloverleaf representation of the primary and secondary structure of $tRNA^{Phe}$. Secondary structure hydrogen bonds are represented by solid bars except for the bonds between G4 and U69, which are indicated by a broken bar, since this is the only basepair connected by non-Watson–Crick hydrogen bonds.

1975). Other tRNA's whose crystal structures have been solved include $tRNA^{Asp}$ (Moras *et al.*, 1980), $tRNA^{Gly}$ (Wright *et al.*, 1979) and the initiator $tRNA^{fMet}$ from *E. coli* (Schevitz *et al.*, 1979) and from yeast (Woo, Roe & Rich, 1980). Since much of the work to be described in subsequent chapters relates to dynamics in $tRNA^{Phe}$, the basic elements of the structure of that molecule will be briefly reviewed here.

The high resolution crystallographic structure of yeast $tRNA^{Phe}$ (Hingerty, Brown & Jack, 1978; Rich, 1978) shows an L-shaped molecule, with each arm being a short segment of double helical RNA. The double helical structure reflects the extensive intramolecular hydrogen bonding of short complementary sections within the molecule. This secondary structure almost certainly occurs in all tRNA's. It was first predicted by

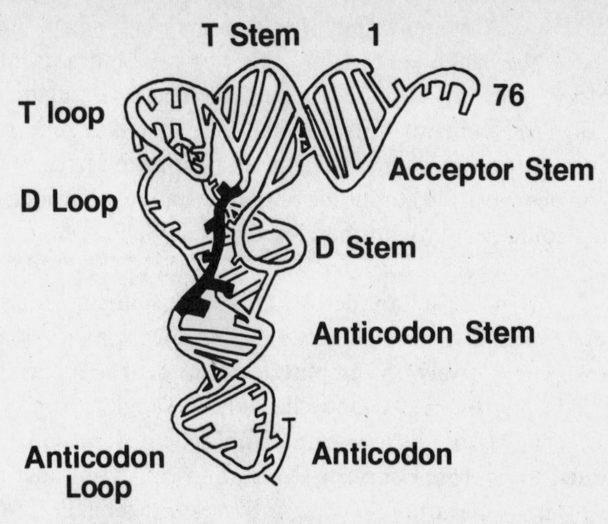

Fig. 2.10. Schematic representation of the secondary and tertiary structure of yeast $tRNA^{Phe}$. The backbone is represented by a continuous tube, while the hydrogen bonding pattern of the secondary structure is represented as bars. The principal structural features identified in figure 2.9 are labeled here, with the exception of the variable loop, where the backbone is shaded.

Holley *et al.* (1965) for $tRNA^{Ala}$, the first tRNA to have its primary structure solved. Certain features of sequence are preserved in all known tRNA structures, and these are summarized in the standard cloverleaf diagram shown in figure 2.9. The three dimensional folding of $tRNA^{Phe}$ is shown in figure 2.10. In addition to the aforementioned double helical regions, the molecule possesses a short single stranded region at the 3′ acceptor terminus, a hairpin loop at the bottom of the anticodon stem, and a compact, complicated structure in the elbow of the molecule where the variable loop, the D loop, and the T loop interact. The combination of such a diverse range of structural features in a single molecule, along with the availability of a large body of experimental evidence from biophysical and biochemical studies, makes $tRNA^{Phe}$ an ideal system in which to study nucleic acid structure and dynamics.

One other conformational feature of individual nucleotides bears describing here. The five-membered furanose ring will not generally have a planar configuration, because this would strain the bond angles at each atom from their ideal values, and because it would also place each of the torsional angles within the ring in the energetically unfavorable *cis* conformation. More favorable conformations can be achieved by puckering the ring into a nonplanar configuration. Since the molecular backbone includes the C3′—C4′ bond within the furanose ring, sugar pucker is closely related to the backbone torsion angle δ and is an important factor in the conformation of the backbone. To describe sugar pucker mathematically, two functions are needed. The first is a function that describes which part of the ring has the greatest deviation from planarity, and the second describes the amplitude of that deviation. Considering that a perturbation could pass around this closed ring in cyclic fashion, the first function is a pucker phase angle, P. The puckering amplitude can be defined as the root mean square deviation of the atomic positions from a suitable reference plane, in which case it is denoted q and is measured in nanometers. Alternatively, the amplitude can be defined in terms of the maximum deviation of the endocyclic torsion angles from the *cis* configuration ($\theta = 0$), in which case the amplitude is denoted θ_m and is measured in degrees. More complete descriptions of these functions are given in the primary literature (Altona & Sundaralingam, 1972; Cremer & Pople, 1975; Rao, Westhof & Sundaralingam, 1981) and in the discussion on sugar pucker dynamics in tRNA (section 6.3).

Detailed descriptions of nucleic acid structure can be found in the reviews by Dickerson *et al.* (1982) and by Parry & Baker (1984), and in the texts by Cantor & Schimmel (1980) and by Saenger (1984). Several excellent books review information on the structure and function of transfer RNA (Altman, 1978; Schimmel, Söll & Abelson, 1979; Söll, Abelson & Schimmel, 1980).

2.4 Molecular associations

The structure and energetics of complexes formed by the binding of molecules to proteins or nucleic acids can be largely understood in terms of the types of interactions described in the preceding sections. One new feature is a general decrease in entropy that results from the loss of relative mobility of two molecules upon binding. When two molecules bind tightly, the entropy of translation and rotation of the two molecules is replaced by that of the complex. Because these entropies depend only weakly on molecular size, the translational plus rotational entropy of the complex will be only about half that of the separated molecules. The exact reduction in the entropy depends on the molecules involved, their concentrations, and state variables such as temperature and pressure. For large molecules at 1 M, the reduction is roughly 200 J K^{-1} mol^{-1} with nearly equal contributions from translation and rotation. This entropy decrease corresponds to an increase in free energy of 60 kJ/mol at 300 K. Any reduction of internal mobility of the molecules upon binding will lead to an additional loss of entropy.

To form a stable complex, the unfavorable free energy change due to the loss of translational and rotational entropy must be offset by other factors. If there are solvent molecules or small ions that are tightly bound to the associating species but displaced during complex formation, there will be an entropy gain from the motion of these displaced molecules. Other factors include the types of interaction that stabilize the native structures of proteins and nucleic acids themselves. For example, examination of the subunit interfaces in protein–protein complexes generally shows that the protein atoms are packed quite densely, that hydrophobic sidechains are clustered together, and that hydrogen bonding and ionic groups are associated so as to compensate for the loss of solvent interactions upon formation of the interface.

Protein–DNA complexes are of great intrinsic interest and also provide a good example of the various factors discussed above. A recent X-ray diffraction study of the repressor protein of coliphage 434, bound to a 14 basepair DNA operator region recognized by the repressor, has been described by Anderson, Ptashne & Harrison (1985). Among the key interactions are bridges formed between several protein residues (positively charged or hydrogen bond donors) and negatively charged phosphate groups on the DNA. Some of the alpha helices in the protein are also oriented so that their sizable net dipole moments are optimally disposed with respect to phosphates. The charge–charge interactions probably more than compensate for the loss of solvent interactions with the separated macromolecules (Fersht *et al.*, 1985). The interactions with phosphates help to orient the repressor so that a surface alpha helix fits snugly into the major groove of the DNA; this major groove binding is a common feature of models for protein–DNA interactions (Pabo & Sauer, 1984; Weber & Steitz, 1984; Matthew & Ohlendorf, 1985) and was also observed in the first protein-DNA co-crystal (Frederick *et al.*, 1984). Specificity of binding appears to be determined primarily by complementary hydrogen bonding and steric interactions between helix side chains and the base pairs that occur in the operator sequences. An important general feature of protein binding to nucleic acids is that the binding of each protein displaces a number of salt ions in the surrounding solution; as a result, the binding equilibria are very sensitive to the type and concentration of salt ions present in the system (Mossing & Record, 1985; Record *et al.*, 1986). When the protein is bound to a DNA segment whose sequence is not the one required for site-specific binding, there may be insufficient complementarity for the protein to displace the hydration layer and the bound counterions. The protein is loosely constrained near the DNA and can diffuse along it until the correct sequence for specific binding is found

(Berg, Winter & von Hippel, 1981; Matthew, Weber, Salemme & Richards, 1983; Berg & von Hippel, 1985).

Additional information on the structural aspects of biomolecular association can be found in Banaszak, Birktoft & Barry (1981); Ohlendorf & Matthews (1983); Pabo & Sauer (1984); Fersht (1985); and Record *et al.* (1986).

3

Dynamics of proteins, nucleic acids, and their solvent surroundings

In this chapter, we provide an overview of the dynamics of water, aqueous solutions, proteins and nucleic acids. Our intent is to provide some sense of the relationships among the different types of motion that will be separately examined in later chapters. We also review a number of basic concepts that will be used extensively in what follows. As with the preceding chapter, this general discussion is not complete. Additional details will be introduced where appropriate in other chapters.

3.1 Water and aqueous solutions

At a temperature of 300 K, liquid water has enough kinetic energy that a typical pair of hydrogen bonded molecules will separate and form new bonds with other neighbors in a time of roughly 4 ps. Smaller displacements of the molecules, corresponding to deformations or transient breaks in hydrogen bonds, occur on a shorter time scale. As a result of this underlying molecular mobility, bulk water behaves as a moderately viscous fluid except at times below about 0.1 ps, where the rigidity of the hydrogen bonded structure becomes apparent.

Solute molecules influence the molecular motion of water in a variety of ways. As described in the preceding chapter, the structure of liquid water adjacent to small hydrophobic solutes is similar to that of bulk water. The mobility of water in hydrophobic hydration shells is slightly reduced in comparison to bulk water, in part because of the reduced number of hydrogen bonding rearrangements available to a water molecule adjacent to the solute. The effect of ions on solvent water mobility can be quite different, depending primarily on the charge density of the ion. For ions with a high charge density, the coordination shell waters adjacent to the ion (and possibly more distant waters) are strongly oriented by the electrostatic field of the ion; the hydrogen bond network of bulk water is broken down and replaced with an induced structure. Such waters are bound relatively strongly to the solute. As examples, the residence times of water molecules in the coordination shells of Li^+, Na^+, and Cl^- are about 30, 10, and 5 ps, respectively. For divalent ions, residence times of 10^{-9} s or more are common. At the interface between the induced and bulk water structures, there is a relatively disordered region of solvent, as mentioned in chapter 2. The mobility of water molecules in this region is slightly greater than that in the bulk.

The mobility of the solute molecules themselves is influenced by all of the factors outlined above. Over very short times (< 0.5 ps), small solutes typically exhibit rattling motions of small amplitude within their solvation cages. Over longer times, rearrangements of the bulk hydrogen bonding network permit more sizable displacements to occur. During such displacements, the motion of the solute is continually interrupted by repulsive core collisions and electrostatic interactions with the fluctuating solvent molecules. Such interruptions are the microscopic basis of the frictional effects that limit the net rate of solute diffusion. The root-mean-square magnitude of solute displacement expected during a time t is

$$\langle r^2 \rangle^{\frac{1}{2}} = (6Dt)^{\frac{1}{2}} \tag{3.1}$$

where the angular brackets indicate an average over different solute molecules, or over different time intervals for a single solute molecule. The quantity D is the diffusion constant of the solute. For spherical molecules, D is given approximately by the equations

$$D = k_B T/f \tag{3.2a}$$

$$f = 6\pi\eta a \tag{3.2b}$$

where $k_B T$ is the Boltzmann constant multiplied by absolute temperature, η is the shear viscosity of the solvent, and a is the solute radius. Equation (3.2a) is the Einstein relationship between D and the friction coefficient f. The friction coefficient connects the drag force $\mathbf{F}$ on a slowly moving solute to its velocity $\mathbf{v}$ relative to the solvent by $\mathbf{F} = -f\mathbf{v}$. The Stokes equation (3.2b) for f is obtained from hydrodynamics for a macroscopic spherical solute, but the equation is approximately valid for small molecule solutes as well. From (3.1) and (3.2), typical displacements of a molecule of 0.2 nm radius will be 0.2 nm during 10 ps and 0.6 nm during 100 ps. Discrepancies from (3.2) do arise due to the microscopic details of the solvent—solute interactions, but rarely exceed a factor of two or so. For example, the diffusion constants of Li^+ and similar ions are smaller than what one would expect based on the radii of the bare ions, primarily because the strongly bound coordination shells increase the effective sizes of these ions.

The internal motions of flexible solute molecules in water can also be described in terms of the preceding discussion. Here, the only new feature is the presence of restoring forces (such as those outlined in chapter 2) that limit the relative displacement of different parts of the solute. The dynamics of the internal motion will reflect the simultaneous action of the random forces that drive molecular diffusion and the systematic forces that control solute conformation. These motions can be characterized in more quantitative fashion by use of phenomenological models. Such models have a long history in the analysis of liquids, polymer solutions, and other dense materials. One such model is summarized by the Langevin equation of motion (Chandrasekhar, 1943). For a single coordinate (e.g., the distance of one group from the center of a large solute), this equation takes the form

$$m\frac{d^2x}{dt^2} = F(x) - f\frac{dx}{dt} + R(t) \quad (3.3)$$

Here, m and x are the mass and position of the group, respectively, and t is the time. The term $F(x)$ represents the effective mechanical restoring force (the mean force) acting on the group. The terms $-f\,dx/dt$ and $R(t)$ represent the direct effects of rapidly varying forces acting on the group (e.g., those due to collisions); the first term is the average frictional force due to the relative motion of the group through its surroundings (f is the friction coefficient), and $R(t)$ represents the remaining randomly fluctuating force. This model, and other phenomenological models (e.g., the diffusion equation, and the Kramers and transition state theories for the crossing of energy barriers) will be considered in more detail later. All of these phenomenological models incorporate effective restoring and frictional forces without attempting a detailed description of the microscopic origin of these forces. These models are of great value in the analysis of molecular dynamics simulations. By using such models, one can focus on particular motions, assign definite magnitudes to the various forces involved, and develop simplified physical pictures that help to deepen our understanding of the dynamics. Analysis of molecular dynamics by use of a phenomenological model is often a helpful intermediate step before analysis by use of more detailed microscopic theories. Phenomenological models are also of value as a basis for simplified dynamical simulations. For example, if one has available estimates of the diffusion constants of key particles in an overdamped system (e.g., from (3.2)), then one can generate typical motions of these key particles without having to follow every detail of less interesting particles in their surroundings. Such stochastic dynamics simulation methods will be described more fully in chapter 4.

Additional information on the microscopic dynamics of water and aqueous solutions is given by Eisenberg & Kauzmann (1969); Wolynes (1980); Franks (1983); and Impey, Madden & McDonald (1983).

3.2 Protein dynamics

As is illustrated in figure 2.3, a polypeptide chain consists of a large number of groups linked by covalent bonds that are intrinsically permissive of rotation. The groups linked by such bonds are themselves comparatively rigid and constitute the fundamental dynamical elements in a protein molecule. Examples of such groups are the CONH peptide groups that link successive residues, and the ring in the tyrosine sidechain (figure 2.3). The typical thermal motions observed within a protein are dominated by the torsional oscillations of these groups about the single bonds that link them together. Within such groups, only small atomic displacements occur due to the large energy cost of deforming bond lengths, bond angles, and dihedral angles about multiple bonds. Superposition of the rigid group displacements yields a remarkably rich dynamical spectrum that ranges from the rapid local motions of the individual groups to slow collective distortions of large regions within the molecule. Some of the motions that occur are indicated in table 3.1; these motions will be considered in more detail in what follows.

Because of the high packing density in protein molecules, their atomic motion displays certain similarities to that seen in other dense materials. Over short periods of time (less than 0.5 ps), the small amplitude motions of the residues display similarities to the motions of molecules in a liquid. Each group is temporarily trapped, rattling in a cage that consists of other groups in the protein and (at the surface of the protein) molecules of the surrounding solvent. The cage atoms frequently collide with the encaged group, rapidly randomizing its motion. For many processes with longer characteristic times, types of motion that are typical of solid materials appear. Such motions are expected, because the atoms of the protein have definite average positions corresponding to the native molecular structure. Because of the interactions that maintain this structure, the protein matrix displays only limited compliance in the larger deformations that occur at long times. The solid-like aspects of protein behavior appear on both the local and global scales. Local group motions of large amplitude are typically opposed by substantial restoring forces associated with the distortion of the cage surrounding the group. Consider, for example, the rotational isomerization of a tyrosine ring (figure 2.3) corresponding to a 180° change in χ^2. If the ring is located in the interior of a protein, a number of protein atoms are located in the volume that would be swept out by

Table 3.1. *Typical features of some internal motions of proteins and nucleic acids*

Motion	Spatial extent (nm)	Amplitude (nm)	Log_{10} of characteristic time (s)
Relative vibration of bonded atoms	0.2 to 0.5	0.001 to 0.01	−14 to −13
Longitudinal motions of bases in double helices (nucleic acids)	0.5	0.01	−14 to −13
Lateral motions of bases in double helices (nucleic acids)	0.5	0.1	−13 to −12
Global stretching (nucleic acids)	1 to 30	0.03 to 0.3	−13 to −11
Global twisting (nucleic acids)	1 to 30	0.1 to 1.0	−13 to −11
Elastic vibration of globular region	1 to 2	0.005 to 0.05	−12 to −11
Sugar repuckering (nucleic acids)	0.5	0.2	−12 to −9
Rotation of sidechains at surface (protein)	0.5 to 1	0.5 to 1	−11 to −10
Torsional libration of buried groups	0.5 to 1	0.05	−11 to −9
Relative motion of different globular regions (hinge bending)	1 to 2	0.1 to 0.5	−11 to −7
Global bending (nucleic acids)	10 to 100	5 to 20	−10 to −7
Rotation of medium-sized sidechains in interior (protein)	0.5	0.5	−4 to 0
Allosteric transitions	0.5 to 4	0.1 to 0.5	−5 to 0
Local denaturation	0.5 to 1	0.5 to 1	−5 to +1

the ring during this rotation. The necessary displacement of these cage atoms associated with ring rotation produces substantial stress in the protein matrix; thus, there is a large energy barrier to the rotation of the ring. The underlying mechanism is similar to that which obtains for vacancy diffusion in crystals, i.e., hopping from one site to another with a strained intermediate state. In certain large scale motions, the distortion of the protein is distributed over many residues and the relative displacements of neighboring atoms are small. The protein can then be pictured as behaving somewhat like a continuous, elastic material. For a globular protein, one simple example would be a 'breathing' motion analogous to the fundamental mode of radial oscillation of a sphere (de Gennes & Papoular, 1969; Suezaki & Gō, 1975). Another example is the hinge bending motion that occurs in proteins having two globular regions that are linked by a region having a smaller cross section. In these types of motion, the protein may display a simple Hooke's-law character. Such motions typically display significant damping due to the solvent surrounding the protein.

In addition to the local structural transitions and global elastic motions, proteins undergo more complicated global transitions on long time scales; these transitions have been compared to structural rearrangements that occur in glassy materials (Goldanskii, Krupyanskii & Flerov, 1984; McCammon, 1984; Ansari *et al.*, 1985; Stein, 1985). The native conformation of a protein comprises a large number of slightly different stable structures that correspond to different local minima in the potential energy surface of the system. Transitions among some of these substates are of direct biological importance. For example, the population of enzyme substates that have relatively high activity may be increased by the binding of a regulatory ligand. Many substates, however, seem to exist as an accidental consequence of the composition of proteins; the extensive noncovalent interactions that stabilize the native conformation are tolerant of slight differences in how groups are packed together. The 'accidental' substates may be of indirect importance in biological function. For example, the redistribution of substate populations upon binding of a regulatory ligand may be kinetically coupled to transitions among the accidental substates. Within a given protein, the energy barriers for such transitions appear to span a large range, from zero to tens of kJ/mol. The complicated, global character of many of these transitions is due to the dense packing of the residues in a protein. Rearrangement of packing in one region may depend on small rearrangements in neighboring regions, so that the transitions have a cooperative character. Good discussions of these issues have been provided recently by Ansari *et al.* (1985) and by Stein (1985).

As has been mentioned, both frictional forces and mechanical restoring forces play roles in the dynamics of proteins. The relative importance of these forces depends on the particular physical process considered. For local oscillations about a stable conformation, both underdamped and overdamped motions occur. Examples include underdamped vibrations of covalent bonds (a result of the large restoring forces together with the small size and correspondingly small frictional effects of the structural units involved), nearly critically damped oscillations of tyrosine rings relative to their cages (McCammon, Wolynes & Karplus, 1979), and overdamped collective distortions of hydrophobic clusters within a protein. Global distortions of the hinge bending type are typically overdamped because of the large surface displacements and solvent damping involved. A similar spectrum of behavior obtains for structural transitions from one stable conformation to another. The rotations of tyrosine rings over the large barriers imposed by the protein matrix are of predominantly inertial character, although frictional effects reduce the rate of such transitions (Northrup *et al.*, 1982*a*). Frictional effects are often dominant in transitions that involve larger groups or smaller energy barriers. Simple examples include the rotation of sidechains at the protein surface (where the substantial restoring forces associated with a protein matrix are absent) and the unwinding of regions of the polypeptide chain from the protein surface (local denaturation).

3.3 Nucleic acid dynamics

When the structural dynamics of nucleic acids are compared with those of proteins, subtle differences are observed, but the overall comparison can best be summarized by paraphrasing a statement attributed to Jacques Monod ('What is true for *E. coli* is also true for elephants, only more so.'). For both macromolecular systems, the most rapid motions are of very small amplitude. Groups of atoms, such as individual bases, move with a rapid, jittering motion, rattling in cages defined by effective local potentials that change relatively slowly. The motions of largest amplitude are those that involve the collective displacement of many atoms. These motions can frequently be identified by doing a normal mode analysis of an all-atom model of the molecule, or by invoking an elastic continuum model for the molecule. These large, slow motions are generally overdamped because of the effects of solvent viscosity, and they consequently have a diffusive rather than periodic nature. As was described in the previous section, it is frequently convenient to think of the motions of atoms about their average positions as being liquid-like when they are of small amplitude and rapid, because the atomic velocities are frequently randomized by collisions with neighboring groups that form temporary cages around them. By comparison, the larger amplitude, slower motions are frequently described in terms of the deformation of a continuous elastic solid, because the atoms and groups of atoms have definite average positions, and substantial restoring forces oppose their motions away from these average positions.

Between the extremes mentioned above, there are the collective motions of groups of atoms whose time scale is intrinsically short, but which occur very infrequently because of steric restrictions. Examples of these motions include the rotational isomerization of backbone atoms and the repuckering of furanose rings. In the case of sugar puckering, for example, a given furanose may spend a relatively long period of time oscillating about a local minimum energy configuration. The activation energy for this process is typically on the order of 10 kJ/mol, reflecting the strain in bond lengths and bond angles that is necessary for repuckering the sugar. The average rate of sugar repuckering is consequently not determined by the time it actually takes to move the atoms from one sugar pucker conformation to another, since this requires only about one ps (Singh, Weiner & Kollman, 1985; Harvey & Prabhakaran, 1986). The macroscopic kinetics are instead dictated by the fact that the accumulation within the ring of the energy necessary for overcoming that barrier is a very rare event.

Overall, then, many of the concepts used in the previous section to describe protein motions can be logically applied to intramolecular motions in nucleic acids as well. For highly localized motions, the differences between the two macromolecular systems are due to their fundamentally different local structures – α helices, β sheets and loops for proteins, and the extensive stacking of aromatic rings in nucleic acids. When one considers large scale motions that extend over regions of several nm, two kinds of differences are anticipated. The first of these will be due to electrostatic effects. Proteins generally carry small net electrostatic charges, whereas nucleic acids are polyelectrolytes whose large net charge must be offset by counterions; the interactions between nucleic acids and the solvent are therefore expected to be quite different than in the case of proteins. The second difference arises from the very extended shape of large nucleic acids. Only in rare cases can one expect there to be similarities between the kinds of motions observed in globular proteins and those observed in nucleic acids; hinge bending motion of the globular domains of tRNA is one possible case. More generally, one would expect nucleic acid motions to resemble those of fibrous proteins. Short segments of these molecules can be treated as elastic rods, while longer pieces can be modeled as wormlike coils. Even here, however, fundamental differences in behavior can be expected. A nucleic acid is essentially a double helical backbone wrapped around the outside of a stack of aromatic bases. Fibrous proteins, on the other hand, do not have the same kind of core. Collagen for example, is a triple helical molecule with the backbone on the inside. Other fibrous proteins – for example, tropomyosin — are essentially 100% α-helical, with the α helix coiled to give a superhelix. Given the differences in underlying structure, it is to be expected that the normal modes of motion and the elastic properties of fibrous proteins may be quite different from those of nucleic acids.

There are a number of large scale motions in nucleic acids where our level of understanding is still very limited, and the models for some of these will be reviewed in subsequent chapters. These include the melting of Watson–Crick hydrogen bonds, both transiently (for proton exchange) and permanently (for denaturation); the structure and motions of single stranded nucleic acids in solution; supercoiling; and the unwinding of the right-handed B-DNA double helix, followed by the formation of the left-handed Z-DNA.

3.4 Molecular association dynamics

The first step in biomolecular association is the diffusional encounter of a pair of molecules (Berg & von Hippel, 1985; Dickinson, 1985; Keizer, 1985; McCammon *et al.*, 1986*b*). The rate of diffusional encounter determines the overall rate of a number of biological processes. For example, a number of enzymes catalyze reactions of their substrates at diffusion controlled rates. Such enzymes are said to have been perfected by evolution. Because the net catalytic rate is already limited by the diffusional transport of substrates to the active site of such an enzyme, there is no selective advantage associated with increased reactivity in the enzyme–substrate complex (Knowles & Albery, 1977).

The frequency of diffusional encounter clearly depends on the mobility or diffusion constants of the individual molecules. For the association of molecules A and B, one is really concerned with the *relative* diffusion constant. Neglecting interactions between A and B, the relative diffusion constant is simply $D_{\mathrm{rel}} = D_{\mathrm{A}} + D_{\mathrm{B}}$, where D_{A} and D_{B} are the diffusion constants of A and B, respectively. Interactions between A and B can increase or decrease the rate of approach of the two molecules. One effect that is always present and that always slows the rate of approach is hydrodynamic interaction. As one of the molecules moves, it tends to move the surrounding solvent. The solvent motion in turn displaces the other molecule. The resulting hydrodynamic interaction can be expressed in terms of a relative diffusion coefficient $D_{\mathrm{rel}}(r)$ that decreases with decreasing distance r between the associating molecules. For simple proteins in dilute solution, hydrodynamic interactions may reduce the encounter frequency by 30% or so (Wolynes & McCammon, 1977). More important are the direct forces (e.g., Coulombic interactions) that may act between A and B. If A and B are oppositely charged, their frequency of diffusional encounter will be increased. Noncentrosymmetric charge distributions may tend to draw ligands specifically toward receptor binding sites, or rotate ligands into productive binding orientations as they approach a receptor. Because direct forces may change association rates by several orders of magnitude, it can be said that molecular recognition really begins with diffusional encounter.

The initial diffusional encounter complex is typically a transient structure that must rearrange to form a more stable complex. In many cases, this rearrangement will involve conformational changes in the receptor and/or in the ligand. Enzymes such as lysozyme, for example, have substrate binding sites that are located in clefts whose widths fluctuate randomly as part of the normal thermal motion of the molecule. Entry of substrate molecules into such binding sites must be preceded by or concerted with opening of the active site cleft. Similarly, the intercalation of certain drug molecules in DNA requires the unstacking of adjacent base pairs in the double helix (e.g., Chaires, Dattagupta & Crothers, 1985), and the reaction of certain carcinogens with DNA requires partial unstacking of base pairs (e.g., Daune, Westhof, Koffel-Schwartz & Fuchs, 1985). If the local binding step proceeds rapidly when suitable conformations have been attained, one can think of the association as a 'gated' diffusion-controlled process (McCammon & Northrup, 1981). Random internal motions determine whether the binding site is available and the ligand has a complementary shape (i.e., whether the reactive gate is open) during a given diffusional encounter between ligand and receptor.

At a more detailed level, the dynamics of molecular association will be influenced by a variety of local motions. Rearrangements of the solvation shells of ligand and receptor are likely to be especially important. For example, as noted in section 3.1, the residence time of a water molecule in the first shell around certain divalent ions is 1 ns or more. Thus, the stripping of solvent alone may result in slow association kinetics for some systems. In polyelectrolytes such as DNA, the rate of binding of large ligands (e.g., RNA polymerase) can be very sensitive to salt concentration. These rate effects arise from the displacement of ions specifically associated with the polyelectrolyte and are much larger than the effects arising from simple ionic screening of Coulombic interactions between the ligand and polyelectrolyte (Record *et al.*, 1986). In addition to such intrinsic effects, the sterically congested environment of many binding sites is likely to retard solvent rearrangement.

4

Theoretical methods

4.1 Survey of approaches

Given the structure of a protein or nucleic acid (e.g., from X-ray diffraction analysis) and a potential energy function, there are a variety of methods that can be used to study the dynamics of the molecule. In the present section, we mention a number of these methods and briefly indicate their strengths and weaknesses. More detailed descriptions of the potential functions and of several particularly important dynamics methods are given in the following sections. A useful, brief summary of some of these methods has been presented by van Gunsteren & Berendsen (1985).

The simplest method for studying motion in biopolymers is essentially static in nature and involves the characterization of low energy paths for specific motions. This is the method of *adiabatic mapping* (section 4.5). In this method, one induces a proposed structural change by forcing the primary atoms involved to move along a given path. The remaining atoms are allowed to move so as to reduce or minimize the potential energy of the whole system at each point on the path. These energies are taken to approximate the change in average potential energy that would occur during a real, spontaneous motion, because the shifts in atomic positions during relaxation correspond roughly to structural fluctuations that would facilitate the motion. The method is easy to use, requires only modest computational power, and has been applied to study both local and large scale structural changes. It does not provide any direct information on time scales or dynamic mechanisms, but approximate results can be obtained by using the relaxed energies in analytic models for the dynamics (e.g., the Langevin equation). Other important limitations of this method are the following: The results are dependent on the path assumed for the motion at the outset. If the motion of interest actually proceeds via a different path, misleading results can be obtained. Secondly, the method ignores certain thermal effects. For example, neither the entropy nor the temperature dependence of the enthalpy are ordinarily obtained, even though these are generally important to a full analysis of the kinetics of the structural change. Finally, quantitative errors can arise as a result of incomplete relaxation. For example, the inefficiency of energy minimization algorithms in relaxing delocalized stresses can lead to overestimates of enthalpy barriers for local structural transitions (cf. section 4.5).

Turning to manifestly dynamic approaches, two methods have been widely used to study atomic motion in biopolymers. The first method is *molecular dynamics simulation*, in which the classical equations of motion for the atoms in the system are solved by numerical techniques for time intervals in the picosecond to nanosecond range (section 4.7). The second method is *normal mode analysis*, in which the motion is described as a superposition of harmonic vibrations whose characteristics are determined by the shape of the potential surface near an energy minimum (section 4.6). The normal mode approach has the advantage that, once the modes of vibration are determined, many time average and dynamic properties can be computed easily by analytic techniques. Also, the separation of actual motions into their principal normal mode components may be helpful in the analysis of protein dynamics. The normal mode approach has several major disadvantages relative to the molecular dynamics approach, however. The matrix eigenvalue calculations that are required for full determination of the normal mode displacements and frequencies are not possible on existing computers for molecules having more than a few hundred atoms. This difficulty can be circumvented by simplifying the system or using special techniques to extract selected normal modes. A more fundamental difficulty is that anharmonic effects are known to be very important in macromolecular dynamics at room temperature (cf. chapter 5). That is, the system moves on regions of the potential surface that are poorly approximated by parabolic extrapolations from the region of any one local minimum. This limitation can be partially overcome by the use of quasiharmonic models (Levy, Srinivasan, Olson & McCammon, 1984*c*). Here, an effective or free energy surface is determined from the atomic displacements observed during a dynamic simulation. The quasi-normal modes then are those for a quadratic approximation to the free energy (the potential of mean force) instead of the potential energy. Such models yield an improved description of large amplitude motions (for which the harmonic potentials tend to be too restrictive), but still do not recognize explicitly the existence of multiple minima in the true potential energy surface, for example. Ultimately, perhaps the greatest difficulty with the normal mode approach is that it can not allow a detailed treatment of solvation phenomena. Although damping effects can in principle be included in a normal mode approach, the rearrangements and diffusional motions of the actual solvent molecules can not be.

The primary limitation of the standard molecular dynamics simulation method is that it is restricted to rather short times (generally less than 1 ns on currently available computers). However, the method can be extended to study phenomena that are intrinsically fast, but that have long apparent time scales because they occur rarely. For example, many reactions and structural transitions have long characteristic times because they involve one or more activated processes. In such a process, the system must surmount an energy barrier that separates stable initial and final states. The barrier crossing process itself is often quite rapid, but the time required for random thermal fluctuations within the system to produce the local atomic momenta required for barrier crossing may be on the order of milliseconds or longer. Processes of this type can be simulated by a two step procedure, the method of *activated molecular dynamics* (section 4.9). First, the free energy barrier is calculated from a sequence of simulations that are of the conventional type, except that the system is constrained to move only within successive regions along the transition path in successive simulations (section 4.8). Second, trajectories are computed starting at the peak of the calculated energy barrier, thereby avoiding the long activation period. Such simulations provide rate constants and a wealth of detailed mechanistic information. The primary limitation of the method is that, as with the adiabatic mapping method, one must have at the outset some idea of the key structural changes involved.

A second approach to increasing the effective length of dynamic simulations is to simplify the model system in some controlled fashion. For example, if one is interested in a localized process (e.g., reaction at an active site), one can separate the system into two parts. One part, which is relatively small, contains the atoms of greatest interest and is described in some detail. The larger part is replaced by a simplified bath model; the effects of the bath appear in the forms of randomly varying forces and systematic frictional and restoring forces that act on the atoms in the detailed part. In other words, the deterministic mechanical surroundings of the detailed part are replaced by a stochastic model with statistical properties similar to those of the original surroundings. Two types of system-bath models have been useful in studies of biopolymer dynamics. In one, molecular dynamics calculations have been performed for clusters of atoms (e.g., at a protein–solvent interface) that are subject to *stochastic boundary conditions* (section 4.7). In the other, *Brownian dynamics calculations* (section 4.10) have been used to simulate the diffusional motion of molecules (e.g., the encounter of substrate and enzyme) that move within a stochastic model solvent. In both cases, longer simulations can be carried out in a given amount of computer time by using bath models in place of detailed models for the surroundings. In the Brownian dynamics simulations, for example, times in the microsecond range are accessible. The primary limitation associated with approaches of this type is that a poorly constructed bath model may lead to artifacts in the remaining detailed model.

Dynamical simulation methods can be used to obtain thermodynamic information on biomolecular systems in addition to descriptions of their time evolution. The use of constrained molecular dynamics simulations to calculate free energy variations along transition paths has already been mentioned. Analogous procedures can be used for nonphysical transitions, such as the creation or transmutation of atoms or groups in a molecule. The methods used for such calculations and their connection to observable properties such as binding constants are described in section 4.8. The structural and thermodynamic properties of a system can also be obtained by Monte Carlo simulations (Valleau & Whittington, 1977). In these, the atoms are moved randomly, but according to probabilistic rules which lead to the generation of a representative variety of structures for given thermodynamic conditions (temperature, volume, etc.). Although this approach has been popular in the study of liquids, it has seldom been used to study detailed models of proteins or nucleic acids. Dynamical simulations have been favored primarily because they directly provide the time evolution of atomic motion. Also, molecular dynamics is far more efficient than traditional Monte Carlo methods in providing equilibrium properties of large molecules (Northrup & McCammon, 1980*a*), although newer Monte Carlo methods may change this situation (see chapter 9).

All of the dynamic simulation methods described above involve the calculation of classical as opposed to quantum trajectories. Many processes of functional interest in biopolymers involve the motion of atoms heavier than hydrogen or the collective motion of groups of atoms at relatively low frequencies; such processes are known to be well described by classical dynamics (Friedman, 1985*a*). Some higher frequency motions such as bond stretching should properly be described by quantum dynamics. Even here, the use of classical dynamics ordinarily has negligible effects on the quantities that are of most interest (e.g, magnitudes of atomic displacements), because the high frequency motions contribute little to the amplitudes of atomic motion and are only weakly coupled energetically to low frequency motions. If necessary, approximate corrections for quantum dynamical effects can be made in structural and thermodynamic data obtained from molecular dynamics simulations. Such corrections can be made conveniently by performing a Fourier analysis of the atomic velocities and applying quantum oscillator formulas to the separate frequency components (Berens, Mackay, White & Wilson, 1983). There are, however, some biological processes in which it is likely that quantum dynamics must be considered in detail. Particularly important are electron transfer reactions (and perhaps some covalent bond rearrangements) where the redistribution of electron density in a transition state may be strongly coupled to the motion of a large number of atoms. Here, the electrons are relatively delocalized and hence sensitive to polarization changes in the surrounding material. Recent developments in techniques for computing quantum path integrals should make it possible to study the coupled electronic and nuclear motion in such systems and, for example, obtain reaction rate constants (Chandler, 1984; Wolynes, 1984). Where electron transfer occurs in an otherwise classical system, it should be possible to use classical molecular dynamics simulation results to construct response functions to be used in the quantum calculations.

4.2 Model functions for potential energy or potential of mean force

For detailed calculations of the flexibility and dynamics of biological molecules, it is essential first of all to know how the potential energy of the molecular system varies with the positions of its constituent atoms. In principle, this potential could be obtained by solving quantum mechanical equations to determine the ground state energy of the electrons and nuclei in the system at each possible set of nuclear positions. The resulting energies form a continuous 'Born–Oppenheimer surface' as a function of the nuclear positions, and this surface describes the potential energy cost for most of the kinds of atomic motion considered here. Quantum calculations are, however, not yet feasible for molecules containing thousands of atoms, especially when their solvent surroundings are also considered. Even if such calculations were feasible, it would be necessary to fit a reasonably simple function of the atomic positions to the quantum results to allow the rapid evaluations of the potential and its gradient that are required in dynamical studies.

In the absence of suitable quantum mechanical functions, a number of different types of model potential energy function have been developed. Ideally, such a function would implicitly include all of the interactions described in chapter 2. In practice, a given function may only include a subset of these interactions and have a correspondingly limited range of applicability. Most theoretical studies of biomolecular dynamics have been based on potential energy functions of the 'molecular mechanics' type

(Burkert & Allinger, 1982). In this approach, one specifies a form for the potential function that is a sum of types of terms often used in the description of simpler systems. A typical model potential function has the form

$$V = (\tfrac{1}{2}) \sum_{\text{bonds}} K_b(b-b_0)^2 + (\tfrac{1}{2}) \sum_{\substack{\text{bond}\\ \text{angles}}} K_\theta(\theta-\theta_0)^2$$

$$+ (\tfrac{1}{2}) \sum_{\substack{\text{dihedral}\\ \text{angles}}} K_\phi[1+\cos(n\phi-\delta)] + \sum_{\substack{\text{nonbonded}\\ \text{pairs}}} \left[\frac{A}{r^{12}} - \frac{C}{r^6} + \frac{q_1 q_2}{Dr}\right] \quad (4.1)$$

The first sum includes a term for every covalent bond in the molecule. Each bond is treated as a simple Hooke's law spring with a characteristic force constant K_b and equilibrium bond length b_0; this is a very good approximation at normal biological temperatures where the bond length fluctuations are quite small. The second term accounts for the deformation energy of valence angles between the covalent bonds to a given atom, and the third term represents the intrinsic deformation energy for twisting about an axis through covalently bonded atoms. Together, these three sums account for variations in the covalent bonding energy of the molecule. The three terms in the remaining sum correspond respectively to the core repulsion, dispersion, and Coulombic interactions between a pair of atoms separated by a distance r. As described in chapter 2, these interactions are particularly important due to the high packing density in proteins and nucleic acids. To conserve computer time, it is conventional to omit calculation of the nonbonded interactions for atoms that are beyond a chosen distance. Gradually scaling the interactions to zero as a pair approaches the cutoff distance is usually preferable to abrupt truncation. Scaling yields atomic distributions that are generally closer to those obtained with the full potential (Brooks, Pettitt & Karplus, 1985) and also minimizes spurious energy fluctuations in dynamical simulations (appendix 3). It should be noted that the last term in (4.1) is actually a potential of mean force or free energy rather than a simple potential energy. This is because the dielectric coefficient D describes the average effect of any charges that are not explicitly included in the model in screening the Coulombic interactions of the explicit charges (see below). The parameters that appear in model potential functions (e.g., the force constants and equilibrium bond lengths) are initially obtained from experimental and quantum mechanical studies of small molecules that are chemically similar to segments of the molecule of interest. These parameters are usually refined to yield correct results (structural, spectroscopic and/or thermodynamic) in calculations on experimentally well-characterized systems before calculations on new systems are attempted.

Early versions of potential energy functions for macromolecules often included an additional term beyond those of (4.1), a term that explicitly represented hydrogen bond energies. This is less common, as it is generally agreed that hydrogen bonds are largely electrostatic in nature (Pauling, 1960) and can be adequately represented by judicious choice of the partial electrostatic charges and Lennard–Jones parameters A and C for the donor atom, the hydrogen atom, and atoms in the acceptor group. An explicit hydrogen bonding term can be added to (4.1) if it is desirable to stabilize particular hydrogen bonds during large scale deformations of the model (Prabhakaran & Harvey, 1985). To conserve computer resources, hydrogen atoms are sometimes included implicitly by adjusting the parameters of the heavy atoms to which they are bonded; the latter are then referred to as 'extended' atoms. If explicit hydrogen atoms are not used, and if the extended donor and acceptor atoms have partial charges of the same sign, an explicit hydrogen bonding term is necessary to overcome the electrostatic repulsion. In this case, the parameters of the hydrogen bonding term must be chosen so that the sum of the nonbonded and hydrogen bonding terms gives a well of appropriate depth (Gelin, 1976; Harvey, Prabhakaran & McCammon, 1985*a*).

Model potential energy functions such as that described above have several advantages that result from their simple functional forms. First, the energy can be computed very rapidly for a given configuration of atoms. This makes possible a determination of the relative stability of different possible molecular conformations. Second, one can readily obtain analytic expressions for the spatial derivatives of the energy. Thus, the forces acting on the atoms and the gradients of these forces can also be computed very rapidly. Finally, because the forms of the various terms have obvious physical interpretations, the potential energy can be analyzed using simple structural concepts. These advantages have led to the use of similar model potential functions in the analysis of vibrational spectra and conformations of small molecules (Wilson, Decius & Cross, 1955; Burkert & Allinger, 1982) and in the computer simulation of atomic and molecular liquids (Wood, 1979).

Equation (4.1) can be made to include solvation effects in two different ways. More accurate, but also more demanding of computer resources, is the explicit incorporation of solvent molecules in the basic model (Hagler & Moult, 1978; Clementi, Corongiu, Jönsson & Romano, 1979; Rossky & Karplus, 1979; Hagler, Osguthorpe & Robson, 1980; Clementi & Corongiu, 1981; Beveridge *et al.*, 1984; Hermans, Berendsen, van Gunsteren & Postma, 1984; Jorgensen & Swensen, 1985; Kim & Clementi, 1985; Mezei, Mehrotra & Beveridge, 1985). The water molecules are then treated on the same level as groups within the biopolymer and are represented by terms in (4.1). Detailed models of this kind have been referred to as 'Born–Oppenheimer level' models of solutions, since both the solute and solvent atoms move under the influence of electron-averaged forces (Friedman, 1981). In such models, the dielectric coefficient D is assigned a relatively small value (in the range 1 to 3) because the major dielectric effect in aqueous systems, namely reorientation of solvent dipoles, is an intrinsic part of the model. The residual dielectric effect, namely polarization of the molecular electron clouds, may be approximated either by use of a D that is slightly larger than one, or by use of $D = 1$ if appropriate effective charges are employed. This explicit approach to solvation allows detailed analyses of structure and dynamics at the biopolymer surface, but has been utilized in full dynamical simulations of proteins and nucleic acids only recently as the necessary computational resources have become available (van Gunsteren *et al.*, 1983; van Gunsteren & Berendsen, 1984; Krüger, Strassburger, Wollmer & van Gunsteren, 1985; Seibel, Singh & Kollman, 1986; Wong & McCammon, 1986*a*, *b*).

The traditional approach to solvation effects in studies of biopolymer dynamics has been to approximate these effects implicitly in (4.1). The only terms that then appear in (4.1) are those associated with the atoms of the biopolymer. The interaction parameters are modified in an attempt to make (4.1) a solvent-averaged potential of mean force (Lifson & Oppenheim, 1960). Among the parameter adjustments that have been used are the following. For the Coulombic interaction between two atoms, the dielectric coefficient D has been set equal to the magnitude of the separation of the atoms expressed in Ångstrom units (1 Å = 0.1 nm) (Gelin, 1976; McCammon *et al.*, 1979). Physically, this corresponds to the tendency of the field lines connecting interior atoms to spread into the high dielectric solvent region as the atoms are drawn apart. In some calculations, this modification has been supplemented by a reduction of the charges of atoms near the biopolymer surface because the field lines between such atoms always extend into the solvent region (Northrup *et al.*, 1981) or, in nucleic acids, because of counterion association with the polymer (Harvey & McCammon, 1981). Another modification that has occasionally been used is the elimination of the attractive dispersion energy for nonbonded interactions that involve a hydrophilic group; the remaining net attraction between hydrophobic groups reflects the expected tendency of such groups to aggregate in water (Mao, Pear, McCammon & Quiocho, 1982*b*). Some effort has been devoted to the development of 'solvation shell' models that may allow a more detailed description of this type of effect (Paterson *et al.*, 1981). Analytic expressions for solvent-excluded volumes and solvent-accessible surface areas have been developed that could be used to approximate hydration energies; since these expressions are differentiable with respect to atomic positions, they are suitable for energy minimization and other molecular modeling studies (Richmond, 1984; Connolly, 1985). One modification of this approach is that of Eisenberg & McLachlan (1986), in which the solvation energy of each atom is calculated as the product of the solvent accessible surface area and a solvation parameter.

Most of the electrostatic models described above are only qualitatively reliable at best. Improvements in the implicit representation of electrostatic interactions of biopolymers are likely to emerge from ongoing studies of several kinds (Berendsen, 1985*a*). For example, solutions of electrostatic boundary value problems for charges embedded in a low dielectric volume (representing a protein or nucleic acid) surrounded by a high dielectric medium (representing solvent water) should yield more accurate descriptions of effective Coulombic interactions that could be incorporated in equations such as (4.1) through appropriate dielectric coefficients (Tanford & Kirkwood, 1957; Shire, Hanania & Gurd, 1974; Warwicker & Watson, 1982; Edmonds, Rogers & Sternberg, 1984; Rogers & Sternberg, 1984; Warshel & Russell, 1984; Gilson, Rashin, Fine & Honig, 1985; Matthew, 1985; Mathew *et al.*, 1985; Shaw, 1985; Zauhar & Morgan, 1985). Extensions of such models to include the effects of small ions dissolved in the solvent (e.g., screening of biopolymer charge interactions, neutralization or ionization of biopolymer groups) also appear possible, as is indicated in some of the above references and in others that are primarily concerned with salt effects (Manning, 1979; Bacquet & Rossky, 1984; LeBret & Zimm, 1984; Matthew & Richards, 1984; Pack & Klein, 1984; Soumpasis, 1984; Frank-Kamenetskii, Lukashin & Anshelevich, 1985; Mills, Anderson & Record, 1985; Murthy, Bacquet & Rossky, 1985; Record *et al.*, 1986; Bacquet & McCammon, 1986). Such developments will be particularly useful in the study of molecular associations; variations in electrostatic interactions play a more prominent role there than in the small amplitude internal motions of biopolymers because of the greater spatial range of charge displacement. For solutes that are in extensive contact with solvent, such as oligomers or random coil polymers, the most promising routes to solvent averaged potentials of mean force are analyses of simulations that include explicit solvent (Mezei *et al.*, 1985) and the application of integral equation methods (Pettitt & Karplus, 1985). These methods can in principle avoid some of the approximations (e.g., concerning solute shape or the spatial correlation of salt ions) that occur in previous work. More generally, a combination of these approaches will be valuable. For example, the analytic framework necessary to extract position dependent electrostatic screening functions from simulations of macromolecules in water has recently been established (Pettitt, 1986). Such functions could be used as effective dielectric coefficients in subsequent simulations without explicit solvent.

Any model potential energy function will contain approximations that restrict its range of validity. Thus, in planning a study of biomolecular dynamics, it is necessary to choose a potential function that faithfully represents the interactions involved in the specific processes of interest, but which may allow conservation of computing resources by neglecting less relevant interactions. For example, structural fluctuations of nonpolar groups that are buried inside a globular biopolymer can be reasonably described over a few ps by a model in which the solvent surroundings and Coulombic interactions are treated very approximately. The dynamics of such groups over short times will be dominated by collisions between the cores of neighboring atoms that lead to random displacements in the dihedral angles. Over longer times, the approximations in the treatment of solvent effects and Coulombic interactions would lead to global distortions of the biopolymer that would alter the local geometry and collision dynamics. In addition to such controllable approximations, there are of course other approximations, associated with our limited understanding of, or ability to include (e.g., due to finite computational resources), some details of the atomic interactions. Fortunately, estimates of the resulting errors can usually be provided, if necessary by repeating calculations with appropriate variations in the model potential function (McCammon *et al.*, 1979).

Additional discussion and detailed forms of potential energy functions of the molecular mechanics type for biopolymers have been given by Lifson, Hagler & Dauber (1979); Berendsen, Postma, van Gunsteren & Hermans (1981); Brooks *et al.* (1983); Levitt (1983*a*); Hermans *et al.* (1984); Weiner *et al.* (1984); Jorgensen & Swensen (1985); and Weiner, Kollman, Nguyen & Case (1986*a*).

4.3 Relationship between energy minimization and molecular dynamics

Computer programs for energy minimization and for molecular dynamics have many features in common and generally use many of the same subroutines. Before examining each of these algorithms in some detail, it will be useful to consider the basic organization which is common to both of them.

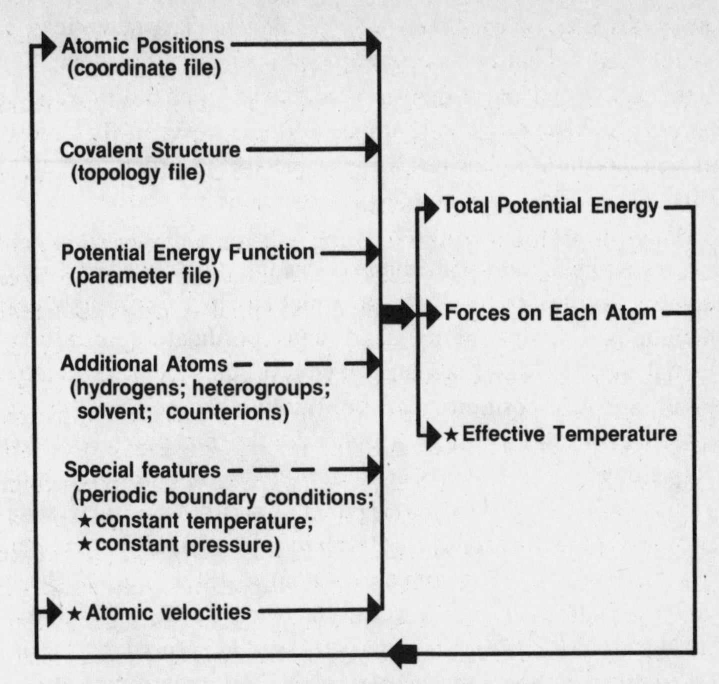

Fig. 4.1. Schematic flow chart of algorithms for energy minimization and molecular dynamics. Features which apply only to molecular dynamics are indicated by asterisks. Dashed lines indicate optional input. Each cycle of energy minimization represents a step in conformation space, while each cycle of molecular dynamics represents a step in time.

Figure 4.1 gives a schematic outline of the energy minimization and molecular dynamics algorithms. At the heart of these programs is a set of subroutines that, given the position of every atom in the model, calculates the total potential energy of the system and the force on each atom. The minimum information needed to run these programs includes a starting set of atomic coordinates, which can be those obtained from X-ray crystallography or any other set of model coordinates, the parameters for the various terms of the potential energy function, and a description of the molecular topology (the covalent structure). The potential energy function is discussed in section 4.2. Since the various covalent interactions are expressed in terms of the internal coordinates (bonds, bond angles, and torsions), it is necessary to specify the covalent structure by identifying the pairs of atoms in all bonds, triplets of atoms defining the bond angles, and quartets of atoms defining torsional rotations about bonds. Provision must also be made for identifying the atoms in hydrogen bonds if an explicit hydrogen bonding term is included in the potential function. Other common features include the use of periodic boundary conditions, so that the simulated system is effectively part of an infinite system; forces of constraint, to hold the system in a particular configuration or move it toward a different target conformation; and provision for occasionally adjusting positions, velocities, or parameters, in order to maintain or change temperature, pressure, volume, forces of constraint, or other externally controlled conditions.

Energy minimization algorithms use the input described in the previous paragraph and the fact that the potential energy function is an explicit, differentiable function of the Cartesian coordinates, to calculate the potential energy and its gradient (which gives the force on each atom) for any set of atomic positions. This information can be used to generate a new set of coordinates in an effort to reduce the total potential energy and, by repeating this process over and over, to optimize the molecular structure. It is thus natural that the first applications of these algorithms to problems of macromolecular structure were to assist in the refinement of model structures of peptides (Scheraga, 1968) and of the crystal structures of proteins (Levitt & Lifson, 1969; Levitt, 1974) and nucleic acids (Jack *et al.*, 1976). More recently, similar procedures to optimize bond lengths and bond angles have become a standard part of crystallographic refinement procedures (Jack & Levitt, 1978; Konnert & Hendrickson, 1980). In addition to these applications to problems of static structure, energy minimization methods can provide some information on dynamic processes. If a model macromolecule is deformed in steps along a chosen reaction coordinate and the rest of the structure is relaxed by minimization while the reaction coordinate is held fixed, the variation of conformational energy along the reaction coordinate gives an estimate of the shape of the energy barrier. This is the basis for the adiabatic mapping method, discussed in section 4.5.

Molecular dynamics is used to animate the molecular model described above. Since one has positions and forces (from which Newton's second law provides accelerations), all that is necessary is that time be treated explicitly and that a suitable set of initial velocities be given to the atoms. Whereas each cycle of energy minimization can be thought of as a step in conformation space, each cycle of molecular dynamics advances time through a very small step, Δt. When this process is repeated many times, the algorithm numerically integrates the equations of motion, providing a solution to the many body problem and producing a simulated trajectory in phase space (i.e., the space defined by atomic position and momentum coordinates).

It should be emphasized that when minimization procedures and the adiabatic mapping method (section 4.5) are used to examine structural transitions, the energies that are calculated are only approximations to the familiar thermodynamic functions. In the absence of thermal motion, there is no entropic contribution to the calculated energy. The latter is therefore more closely related to the thermodynamic internal energy or enthalpy than to a free energy, but does not include the temperature dependent contributions to these functions. For static models, the vibrational entropy and other thermal contributions can be estimated from normal mode analyses (section 4.8). Molecular dynamics, on the other hand, explores the conformational space in a fashion that directly reflects the density of states, so entropic and other thermal effects are naturally included. As a consequence, free energy differences can be determined by molecular dynamics simulations of different conformations along a transition pathway (section 4.8).

4.4 Energy minimization

From a computational viewpoint, the problem of minimizing the energy of a model macromolecular system falls into the general area of nonlinear optimization problems. Given a set of independent variables $\mathbf{x} = (x_1, x_2, x_3, \ldots, x_n)$ and a specified objective function $V = V(\mathbf{x})$, the task is to find that set of values for the independent variables, denoted $\mathbf{x}^*$, for which the function has its minimum value $V(\mathbf{x}^*) = \min(V(\mathbf{x}))$. In the case of a model macromolecule containing N atoms, the $3N$ components of $\mathbf{x}$ are the atomic coordinates and V is the potential energy, calculated from the potential energy function (section 4.2).

In this section we give a brief overview of some of the procedures that have been developed for nonlinear optimization, concentrating on discussions of the two algorithms most commonly applied to energy minimization in macromolecular modeling, namely the steepest descent and conjugate gradient methods. A brief discussion of more powerful methods is given at the end of this section. For the sake of clarity in the discussion, it will occasionally be convenient to cite as examples the one dimensional and two dimensional problems. For the latter we note that two independent variables produce an energy surface that is analogous to a topographical map. In that case, possible independent variables would be latitude and longitude, and finding the minimum is analogous to locating the lowest point in a landlocked country surrounded by a mountainous border (so the lowest point is in an interior valley and not on the frontier).

Before examining the various energy minimization methods, it should be pointed out that it is extremely difficult to find the global minimum of a general nonlinear function with ten or more independent variables. Even molecular systems with as few as ten atoms may have conformational spaces that are so complicated that finding the global minimum is almost impossible. For macromolecules, only an extremely tiny fraction of conformational space can be explored. Energy minimization methods can therefore be used to refine molecular structures in the sense of eliminating the worst steric conflicts and adjusting bond lengths and bond angles to values near their respective optima, and they can provide information on the relative energies of different conformations, but they generally will not produce structures that are very different from the initial, unrefined structure. Exceptions to this rule are discussed at the end of this section.

Any continuous, differentiable function of the independent variable x can be expanded as a Taylor series about the point x_0,

$$f(x) = f(x_0) + (x - x_0) f'(x_0) + (x - x_0)^2 f''(x_0)/2 + \dots \tag{4.2}$$

The generalization of this equation from the one dimensional to the many dimensional case requires the replacement of x by the vector $\mathbf{x}$ and the introduction of matrices for the various derivatives, as described in standard texts (Jacoby, Kowalik & Pizzo, 1972). A given optimization method can be classified by its *order*, which is defined by the highest order derivative that is used in the method. We will briefly discuss one zeroth order method (grid search) and one second order method (Newton–Raphson) and then turn to two first order methods (steepest descent and conjugate gradient).

The conceptually simplest way of locating the minimum of $V(\mathbf{x})$ is to scan the space defined by $\mathbf{x}$ in regular increments of x_1, regular increments of x_2, and so on. For the two dimensional problems, this is equivalent to laying a grid over the surface and picking that grid point at which $V(\mathbf{x})$ is a minimum. The ability of this grid search procedure to locate the global minimum depends on the fineness of the grid and the roughness of the surface. An iterative procedure can be established by beginning with a relatively coarse grid and using successively finer grids to examine the neighborhood of the minimum located by the previous cycle. Even for problems of small dimensionality, grid search methods are only useful for surfaces with nearly monotonic behavior away from the minimum. For surfaces with multiple local minima or for problems with many degrees of freedom, these methods are inefficient and can converge to false minima; accordingly, they are rarely used in energy minimization studies on macromolecules.

The Newton–Raphson method is a second derivative method that is based on the assumption that, in the region of the minimum, the energy depends approximately quadratically on the independent variables. For the one dimensional case, $V(x)$ in the neighborhood of the minimum is assumed to be of the form

$$V(x) = a + bx + cx^2$$

where a, b and c are constants. The first two derivatives are:

$$V'(x) = b + 2cx \tag{4.3}$$

$$V''(x) = 2c \tag{4.4}$$

At the minimum, $V'(x^*) = 0$, and x^* can be calculated from (4.3),

$$x^* = -b/2c \tag{4.5}$$

Substituting (4.3) and (4.4) into (4.5) gives

$$x^* = x - V'(x)/V''(x) \tag{4.6}$$

If the system is at the point x, and if the first two derivatives are known at this point, (4.6) predicts the location of the minimum. This relationship can be generalized to the multidimensional case quite easily (Jacoby *et al.*, 1972), and it serves as the basis for the Newton–Raphson method. For quadratic energy surfaces, no iteration is necessary. For nonquadratic but monotonic surfaces, this method and its modifications have widespread utility. There are three reasons why it has rarely been used in problems of macromolecular energy minimization. First, the highly nonquadratic character of the energy surface and the presence of multiple local minima render the surface unsuitable for examination by the Newton–Raphson algorithm. Most macromolecular applications of the method have been to complete a minimization that was initiated by another method; the first problem then does not arise. Second, the potential energy function (section 4.2) is expressed in terms of a mixture of internal coordinates (e.g., bond lengths and angles) and Cartesian coordinates, and only recently have convenient analytic methods been developed for calculating the second derivative matrix (Brooks *et al.*, 1983; Nguyen & Case, 1985). Third, although the second derivative matrix can be approximated numerically or calculated analytically, it is necessary to invert it (equation 4.6), a very difficult problem for macromolecular systems where the matrix is large ($3N \times 3N$) and frequently ill-conditioned. One modification, adopted basis set Newton–Raphson (ABNR) minimization, is incorporated into the CHARMM package of programs for macromolecular mechanics (Brooks *et al.*, 1983).

First order methods truncate (4.2) after the second term, making use of information about the local slope of the potential energy surface (the first derivative), but not about the curvature (the second derivative). In physical terms, the first derivative represents force, since force is the negative gradient of the potential,

$$\mathbf{F} = -\nabla V \tag{4.7}$$

It is logical to use this information in refining macromolecular structures: the potential energy can be reduced by moving each atom in response to the net force acting on it. Since the potential function in section 4.2 is an explicitly differentiable function of the atomic coordinates, the calculation of the force in (4.7) is straightforward.

The first order methods to be discussed here fall into the general class of iterative *descent techniques* (Jacoby *et al.*, 1972, chap. 5). The molecular configuration prior to the kth iteration is specified by the $3N$ dimensional vector $\mathbf{x}_{k-1}$. Each iteration consists of three parts. First, a descent direction is chosen; this can be represented by a $3N$ dimensional vector of unit length, $\mathbf{s}_k$. Second, a descent step size, specified by the scalar λ_k, is determined. Third, the descent step is taken according to the relationship

$$\mathbf{x}_k = \mathbf{x}_{k-1} + \lambda_k \mathbf{s}_k \tag{4.8}$$

In the steepest descent method, the direction of displacement is parallel to the net force,

$$\mathbf{s}_k = \mathbf{F}/|\mathbf{F}|$$

Since this vector is parallel to the negative gradient of the energy, equation (4.7), it points straight downhill, whence the method derives its name. In the two dimensional case, the pathway defined by successive descent vectors is analogous to the downhill flow of water, so that, in principle, the method leads to the nearest local minimum. The step size is generally determined by a simple criterion. An initial step size is chosen and the first step is taken. If the energy is, in fact, reduced by this step, the step size is increased by some multiplicative factor (typically 1.2) for the next iteration, a process which is repeated as long as each iteration reduces the energy. Whenever an iteration produces an increase in energy, it is assumed that the step size was too large, resulting in a leap across the valley containing the minimum and up the slope on the opposite side, so the step size is reduced by a multiplicative factor (typically 0.5) for the next iteration. This process of iteratively adjusting the step size is intended to make successive descent steps of a size appropriate to the curvature of the region being explored. In our hypothetical two dimensional problem of exploring a country, a step of 10 km or more might be appropriate in a large, flat plain, whereas a step of 1 m or less might be required in a small, twisting gully.

Steepest descent is a conservative algorithm in the sense that it tries to head doggedly downhill without crossing local barriers to search for deeper nearby minima. Physically, it makes very limited perturbations to the starting molecular structure. Since the direction of the gradient is determined by the largest interatomic forces, steepest descent can eliminate the worst steric conflicts and bring bond lengths and bond angles to values near their canonical values, but it will not produce the collective motions (corresponding to weak forces on many atoms) that are necessary to generate optimum overall stereochemical structures. It is a nonconvergent method and is particularly inefficient for multidimensional problems that involve irregular potential surfaces with many local minima, such as those characteristic of macromolecular energy calculations.

A generally more efficient descent technique is the conjugate gradient algorithm. Whereas steepest descent chooses the descent direction based entirely on the gradient at the current step, conjugate gradient combines information on the current gradient with that based on the gradients at previous steps. We will describe this method for the two dimensional case, with only a few remarks about the multidimensional case. More detailed descriptions are available in the original literature (Hestenes & Stieffel, 1952; Fletcher & Reeves, 1964) and in standard texts (Dixon, 1972; Fletcher, 1972; Jacoby *et al.*, 1972; Fletcher, 1980).

The conjugate gradient method is depicted schematically in figure 4.2 for a two dimensional quadratic energy function. Starting from the initial point A, where the gradient is $\mathbf{g}_1$, the first search direction is taken to be along the negative gradient,

$$\mathbf{s}_1 = -\mathbf{g}_1,$$

which is along the line ABD. The minimum value of the energy along this search direction occurs at the point B, where the negative gradient $\mathbf{g}_2$ is along the line BC. Whereas steepest descent would take the next step in that direction, conjugate gradient takes the second search direction to be a linear combination of the current gradient and the previous search direction,

$$\mathbf{s}_k = -\mathbf{g}_k + b_k \mathbf{s}_{k-1} \tag{4.9}$$

where the parameter b_k is a weighting factor equal to the ratio of the squares of the magnitudes of the current and previous gradient,

$$b_k = |g_k|^2/|g_{k-1}|^2$$

It can be proven that, for a quadratic surface, the search direction specified by these last two equations will pass through the minimum on the nth step for an n dimensional surface, as long as the minimum along each successive

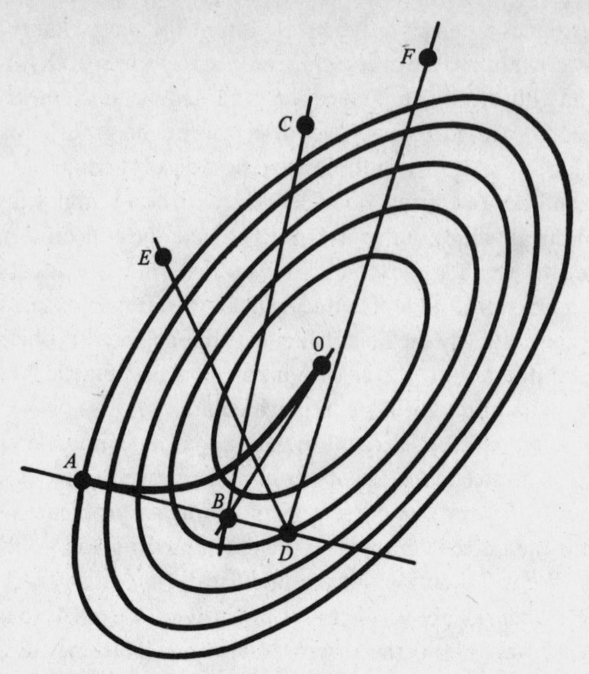

Fig. 4.2. Representation of paths followed by different energy minimization algorithms on a hypothetical two dimensional potential surface. The minimum is at 0, and the ellipses represent contours of potential energy, spaced at equal intervals. Starting from point *A*, the path that would be taken by steepest descent with an infinitesimal step size is identical to that which would be followed by water flowing downhill, represented by the heavy curving line. For larger step sizes, different paths would be followed by steepest descent and conjugate gradient; see text for a more detailed discussion.

search direction is found. Thus, for the two dimensional case in figure 4.2, if the first step goes from *A* to *B*, the second search direction will pass through the minimum at 0.

Even if the step size is not optimum, the conjugate gradient method will still yield a search direction that is superior to that of steepest descent. If, for example, the first step in figure 4.2 were larger than the optimum, arriving at point *D*, steepest descent would follow the gradient in the direction *DE*, while the last two equations would produce a search direction $\mathbf{s}_2$ in the direction *DF*, which passes much nearer the minimum. As a consequence, this algorithm is much more efficient than steepest descent, a fact that is generally true regardless of the dimensionality of the problem or the nature of the potential surface.

One modification of the conjugate gradient algorithm should be mentioned. For the n dimensional case, the method will not converge in n steps unless the potential surface is quadratic. In general it is not; numerical errors can then accumulate in the determination of successive search directions, since the right hand side of (4.9) depends on the previous search direction, $\mathbf{s}_{k-1}$, rather than on the previous gradient, $\mathbf{g}_{k-1}$. To prevent this accumulation of errors, the algorithm is interrupted and reset every m steps, with $m \leqslant n$, by basing the search direction on the current gradient only, setting $b_m = 0$ in (4.9) (Fletcher & Reeves, 1964).

Although algorithms that are still more powerful and efficient are available for nonlinear optimization (Dixon, 1972; Fletcher, 1972; Jacoby *et al.*, 1972; Fletcher, 1980), they have not been widely used in macromolecular energy minimization. In part, this is because steepest descent and conjugate gradient can generally resolve the worst steric conflicts in model structures. This also reflects the fact that even the most powerful algorithms would never be able to find the global minimum for complex systems with hundreds of degrees of freedom and a very large and unknown number of local minima. Since the absolute minimum cannot be reached, anyway, all that is usually required is a method that is capable of eliminating the most serious steric problems and that can provide a measure of the energy difference between two structures. Both steepest descent and conjugate gradient are well suited to this.

As mentioned above, none of the commonly used minimization algorithms generate large collective atomic displacements, since they do not efficiently simulate the transmission of forces through the structure. For problems where such collective displacements might be desired, there are two minimization procedures that can be used. One approach, suitable for the proper folding or packing of one small region during large scale deformations of the molecule, is the pseudodihedral algorithm (Harvey & McCammon, 1982). In this method the molecule is divided into regions by connecting pairs of atoms with virtual bonds (Brant & Flory, 1965; Olson & Flory, 1972; Olson, 1980) which serve as the axes about which rigid rotations of groups of atoms take place. By keeping the number of virtual bonds small — about a dozen or less — it becomes feasible to perform a grid search on the conformational space defined by these pseudodihedral rotations. This local grid search method helps to select an initial conformational point from which to begin serious minimization. In our large scale bending study on tRNA, the combination of the pseudodihedral algorithm with steepest descent was found to reduce the required computation time by about 40 % (Harvey & McCammon, 1982). A second procedure, suitable for very extensive conformational changes such as folding an entire molecule, combines a powerful minimizer, the variable metric method (Fletcher, 1972) with a 'soft atom' potential function. Minimization takes place in the space defined by torsion angles, and the soft atom potential allows atoms to pass through one another. The method was developed by Levitt (1983*c*) for folding model proteins into trial conformations suitable for further refinement by more conventional methods.

Finally, it should be mentioned that molecular dynamics is an alternative to energy minimization for optimizing molecular models; this is discussed in more detail in section 4.7.

4.5 Adiabatic mapping

The simplest application of energy minimization methods to problems of macromolecular dynamics has been the estimation of energy costs for transitions between two or more conformations of a given molecule. If a suitable reaction coordinate can be defined to describe the variation of conformation from one state to another, and if a suitable theoretical model for motion along the reaction coordinate can be found, then the dynamics of the system frequently follow directly from the dependence of conformational energy on the reaction coordinate.

As an elementary example, consider the rotation of a single tyrosine ring through an angle of $\Delta\chi^2 = 180°$ in the interior of a protein. This ring flip is modeled as an activated process (Glasstone, Laidler & Eyring, 1941; Berry, Rice & Ross, 1980). The reaction rate constant is approximately

$$k = (k_B T/h) \exp(-\Delta G_a/k_B T) \quad (4.10)$$

a quantity which can be predicted if the free energy of activation, ΔG_a, is known. To estimate this energy, the tyrosine ring can be rotated in increments of a few degrees, with energy minimization used to relax the steric stresses that are generated. The transition state will correspond to the conformation for which the energy is a maximum, and the difference between this energy and that of the initial conformation is used as an estimate of ΔG_a. In the biomolecular area, energy minimization was first used to estimate the heights of energy barriers in studies of conformational changes of small peptides (Scheraga, 1968). The connection between such barriers and rate constants was first exploited for proteins in a study of tyrosine ring flips in bovine pancreatic trypsin inhibitor by Gelin & Karplus (1975); that study and subsequent refinements are described in chapter 6.

A second class of problems that can be studied by comparing the relative energies of different conformations is the large scale bending of proteins and nucleic acids. If the effective energy profile along the reaction coordinate is nearly quadratic, the restoring force can be calculated and motion along this coordinate can be examined by diffusive models that use the Langevin equation (section 3.1). The first problem to be studied by this method was the hinge bending motion of lysozyme (McCammon, Gelin, Karplus & Wolynes, 1976); that application and the study of other protein systems by the same procedure is described in section 7.2. Levitt (1978) was the first to examine structural deformations in a nucleic acid using energy minimization; that investigation and other nucleic acid studies are described in section 7.3.

The calculation of effective energy costs for transitions by displacement along a reaction coordinate followed by relaxation is referred to as adiabatic mapping. This name derives from an assumption implicit in the use of such energies in discussions of dynamics, namely that the relaxation produced by energy minimization would occur rapidly (adiabatically) on the time scale of the transition during actual dynamics. The resulting energies are perhaps best thought of as rough approximations to the enthalpic component of the potential of mean force. The potential of mean force is a function of the reaction coordinate; for each value of this coordinate, it specifies the free energy of the system corresponding to a thermal average over all coordinates *other* than the reaction coordinate (cf. sections 2.1 and 4.8). If the motion along the other coordinates is faster than that along the reaction coordinate, the potential of mean force is the *true* effective energy for the transition because the systematic force along the reaction coordinate in the fluctuating molecule is the negative gradient of this potential (McQuarrie, 1976; Friedman, 1985*a*). Even when these time scales are not separated, the potential of mean force (and the adiabatic mapping approximation to it) is useful in describing the probability of occurrence of different values of the reaction coordinate (section 6.2). As discussed in section 4.3, the energy obtained by adiabatic mapping is *not* exactly the same as the potential of mean force, however.

The adiabatic mapping procedure is straightforward. The molecular model is relaxed by extensive minimization to produce a starting conformation. It is then deformed in a series of small steps (relative atomic displacements of 0.1 nm or less are typical), with energy minimization after each step to produce structures whose energies are compared to those of the starting conformation. For large deformations requiring many steps, problems arise if the energy minimization methods cannot remove the accumulated stresses efficiently enough to produce structures that are truly comparable to the starting conformation. We have observed, for example, that models of tRNA that have been bent through angles greater than 30° have conformations whose energies converge at different rates under energy minimization, because different regions of the models are at

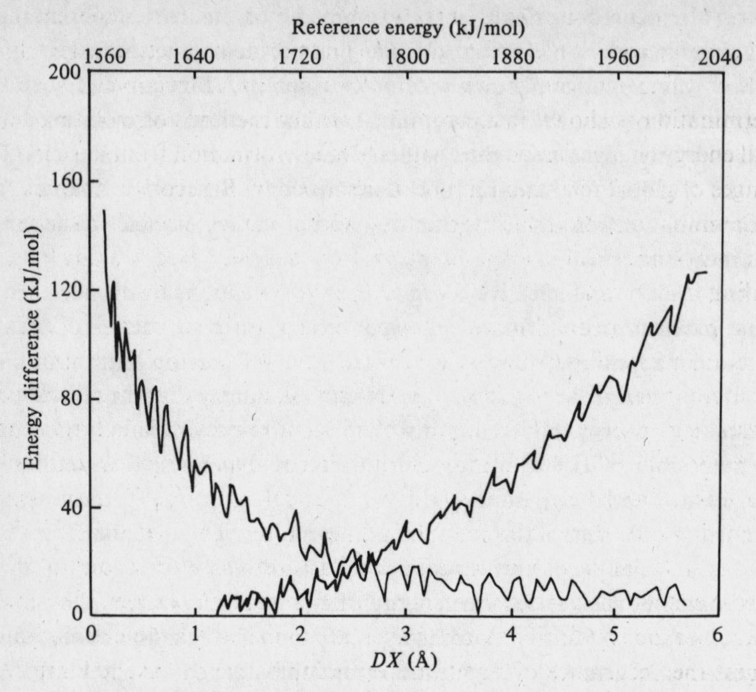

Fig. 4.3. Estimation of the energetic cost for bending a $tRNA^{Phe}$ model, using two methods to monitor the degree of refinement of the structure (Tung *et al.*, 1984). 150 steps of steepest descent minimization are shown, and the energy difference between the bent and unbent structures is plotted for the reference energy method (energy difference *vs* reference energy, curve starting at the upper right and descending to the lower left) and for the atomic deviation method (energy difference *vs* root mean square displacement during refinement, *DX*, curve starting at upper left and descending to the lower right). Both curves reach a plateau, and the last 20 points of each curve are used to estimate the final energy difference. In this particular example, the reference energy method gives 5.4 kJ/mol and the atomic displacement method gives 7.4 kJ/mol. (Copyright © 1984, John Wiley & Sons, Inc. Reprinted by permission.)

drastically different levels of refinement. We have developed two methods for defining the degree of refinement of molecular models (Tung, Harvey & McCammon, 1984). One, the reference energy method, calculates the energy of a reference region far from the site of imposed deformations. The other, the atomic deviation method, calculates the root mean square deviation of the atomic positions from their starting positions for those atoms in the reference region. Either the reference energy or the reference mean square deviation can be used as a parameter to monitor refinement, since the conformation of the reference region is not affected by the imposed deformation. With either of these as an indicator, the energy difference between a deformed and an undeformed structure at similar levels of refinement can be accurately determined. An example of such a determination is shown in figure 4.3. Another method for isolating the small energy changes associated with a given deformation from the energy changes of global relaxation has been described by Bruccoleri, Karplus & McCammon (1986). This approach, which makes use of harmonic constraint forces to control global relaxation, has been used to study hinge bending in lysozyme.

The problem of inadequate refinement also arises in studies of local structural transitions. As discussed in section 4.4, energy minimization algorithms relax large, localized stresses quickly but may relax delocalized stresses (e.g., due to elastic twisting, bending or expansion of a large part of a macromolecule) very slowly. In cases where local structural changes have been studied by adiabatic mapping, different results have been obtained in subsequent studies that made use of more powerful dynamical relaxation techniques. A dramatic example is provided by studies of the barriers encountered by an oxygen molecule in moving between the heme pocket and the exterior of myoglobin. The original adiabatic calculations suggest the occurrence of significant enthalpy barriers (Case & Karplus, 1979), but the dynamic calculations suggest that the enthalpic component of these barriers is quite small at 300 K (Case & McCammon, 1986). Apparently, the results of the original adiabatic calculations reflected slowly relaxing, distributed stresses that were left after relaxation of the initial nonbonded repulsion between ligand and protein. In calculations of the energy cost of creating an intercalation site in DNA, much smaller values were found when molecular dynamics was used instead of energy minimization methods to relax the nucleic acid (Prabhakaran & Harvey, 1985). Comparisons such as these suggest that, in large systems, adiabatic mapping may be most useful for *qualitative* analysis of structural changes and determination of *upper bounds* for the associated energy changes.

4.6 Normal mode analysis

The vibrational modes of molecular motion can be excited by the absorption of infrared light or by scattering of shorter wavelengths with frequency shifts that correspond to energies of infrared photons. Consequently, infrared absorption spectroscopy and Raman scattering are the usual experimental methods for studying molecular vibrations. The interpretation of these experiments rests on the ability to identify which modes of vibration are responsible for the various peaks in the spectra. Normal mode analysis has been widely applied for many years to the study of vibrational motions in small molecules (Wilson *et al.*, 1955). The comparison of the experimentally observed frequencies of Raman and infrared absorption peaks with those calculated for molecular models has traditionally been one of the most stringent tests for the values of the force constants for bond stretching and valence angle bending that are used in molecular mechanics potential functions like those of (4.1).

There are two difficulties in applying these methods to macromolecular systems. First, the lowest frequency modes of large molecules, which represent the largest amplitude collective motions, occur in the far infrared at frequencies below 200 cm^{-1}, where observed peaks are generally broad and weak and Rayleigh scattering begins to dominate. Second, as will be discussed in more detail below, the normal mode analysis of large molecules is a computationally demanding problem because of the large matrices that are involved. The first problem is generally attacked by careful measurements using improved instruments (Brown, Erfurth, Small & Peticolas, 1972; Genzel *et al.*, 1976; Twardowski, 1978; Painter & Mosher, 1979; Painter, Mosher & Rhoads, 1981; Urabe & Tominaga, 1982) or by neutron time-of-flight spectroscopy (Bartunik, Jollès, Berthou & Dianoux, 1982). The second problem can be attacked by reducing the number of degrees of freedom with models in which groups of atoms are joined into virtual extended atoms and simplified approximate potentials are used to describe the interactions between these pseudoatoms (Noguti & Gō, 1982; Levy, Karplus, Kushick & Perahia, 1984*a*; Levy, de la Luz Rojas & Friesner, 1984*b*). Alternatively, one can use special techniques to study selected modes rather than the full spectrum (Harrison, 1984; Brooks & Karplus, 1985). If adequate computer memory is available, the usual normal mode methods can be applied directly to molecules with molecular weights up to several thousand (Brooks & Karplus, 1983; Gō, Noguti & Nishikawa, 1983; Levitt, Sander & Stern, 1983, 1985; Tidor, Irikura, Brooks & Karplus, 1983).

In this section we present a brief, qualitative description of normal mode analysis and some of the issues that arise when the method is applied to large molecules. A good introduction to the field of vibrational modes in macromolecules has been given by Peticolas (1979), and the reader should refer to the original papers cited in the previous paragraph for more detailed discussions of the experimental and theoretical approaches. Rigorous derivations of the relevant equations are available in many texts on classical mechanics, such as Goldstein (1980) and Symon (1960).

Normal mode analysis is the multidimensional treatment of coupled harmonic oscillators. To apply it to molecules, it must be assumed that the atoms are point masses, that the molecular conformation is at a local minimum in conformational energy space, that the potential energy is a quadratic function of the coordinates used to describe the motions, and (in the usual treatments) that interactions with the solvent are negligible. Only the first of these assumptions is, of course, reasonable for macromolecules in solution. The conformational energy space sampled at ambient temperatures has a very complex shape with many minima, the potential of mean force is highly anharmonic (see sections 5.2 and 5.3), and low frequency collective motions are often highly overdamped and diffusive, rather than periodic, in character. Nonetheless, the normal mode method is attractive because of its conceptual simplicity, because exact analytic solutions can be obtained for the equations of motion, and because of its proven success for small molecules. Model thermodynamic properties can also be calculated, again subject to the assumptions listed above. The ultimate utility of the normal mode procedure for macromolecules will naturally depend on its ability to explain experimental results and on its predictive powers.

The equation of motion for a simple one dimensional harmonic oscillator of mass m and restoring force constant k can be written

$$m\ddot{x}+kx = 0 \tag{4.11}$$

where x is the displacement from the minimum energy position, and the double dot is the simplified notation for the second derivative with respect to time. The solution to (4.11) is a periodic one with amplitude A:

$$x(t) = A\sin(\omega t-\delta) \tag{4.12}$$

where δ is a phase angle and the angular frequency is

$$\omega = (k/m)^{\frac{1}{2}}$$

This one dimensional model can be generalized to the multidimensional case for a macromolecule if the assumptions of the previous paragraph are satisfied. The single coordinate x is replaced by a set of coordinates q_i that can be collected into a $3N$-dimensional vector $\mathbf{q}$ (where N is the number of atoms). The mass is replaced by a $3N \times 3N$ matrix $\mathbf{M}$ whose elements are given by

$$M_{ij} = \sum_{k=1}^{N} m_k\left(\frac{\partial \mathbf{r}_k}{\partial q_i}\cdot\frac{\partial \mathbf{r}_k}{\partial q_j}\right) \tag{4.13}$$

In this equation $\mathbf{r}_k$ is a vector specifying the position of atom k. The force constant in (4.11) is also replaced by a $3N \times 3N$ matrix, $\mathbf{K}$, whose elements are the second derivatives of the potential energy V:

$$K_{ij} = (\tfrac{1}{2})\frac{\partial^2 V}{\partial q_i \partial q_j}$$

The multidimensional analogue to (4.11) is, then

$$\mathbf{M}\cdot\ddot{\mathbf{q}}+\mathbf{K}\cdot\mathbf{q}=\mathbf{0} \tag{4.14}$$

Formally, the solution to (4.14) reduces to a series of coordinate transformations that first convert the equation to a form in which it becomes an eigenvalue problem and then solve that problem by expressing the equation in a coordinate system in which all the matrices are diagonal. The matrix equation then reduces to a set of independent linear equations. Each of those equations is similar to a normalized form of (4.11) in which the mass is unity. The solution to this system of equations is, in analogy with (4.12), a set of periodic functions. Six of these, corresponding to the translational and rotational degrees of freedom, have zero frequencies, and the remaining $3N-6$ terms correspond to the normal modes of the system, each of which has a characteristic frequency. Each normal mode is a collective, periodic motion, with all parts of the system moving in phase with each other.

The motion of the macromolecule is the superposition of all the normal modes. For a given temperature, the mean square amplitude of motion in a particular mode is inversely proportional to the effective force constant for that mode, so the motion tends to be dominated by the low frequency vibrations. Moving pictures have been generated that depict the nature of some of these modes in bovine pancreatic trypsin inhibitor (Brooks & Karplus, 1983), and that identify the modes corresponding to segmental motions in four proteins (Levitt *et al.*, 1985). Similar films demonstrate low frequency twisting and bending modes in DNA (Tidor *et al.*, 1983).

Discussions of the results of normal mode analyses of proteins and nucleic acids and a comparison of those results both with molecular dynamics simulations and with experimental results will be found in chapter 5.

4.7 Molecular dynamics

This section gives an overview of molecular dynamics and some of its variations. Additional information can be found in appendix 1, which discusses alternative methods of integrating the equations of motion; appendix 2, which gives a step-by-step description of how a typical simulation is carried out; and appendix 3, which explains the modifications necessary for performing molecular dynamics at constant temperature and/or pressure.

In molecular models that ignore quantum dynamical effects, each atom is considered to be a point mass whose motion is determined by the forces exerted upon it by all the other atoms, as described by the equations of motion of classical mechanics. Although it is frequently a straightforward procedure to write down the equations of motion for systems of interacting particles, classical physics provides closed form solutions to those equations for only some of those systems, and only for cases involving one or two independent particles. For three or more independent particles, the so-called many body problem, the equations of motion must be solved numerically.

In a system of N particles, we indicate the mass of particle i by m_i and the position of the particle by the vector $\mathbf{r}_i$. If we use a dot to denote differentiation with respect to time, the relation between the velocity and the momentum of this particle is

$$\dot{\mathbf{r}}_i = \mathbf{p}_i/m_i \tag{4.15}$$

The equation of motion for the particle is

$$\dot{\mathbf{p}}_i = \mathbf{F}_i \tag{4.16}$$

where $\mathbf{F}_i$ is the net force exerted on particle i by all the other particles in the system. This force is given by the negative gradient of the N particle potential energy function (4.1) with respect to the position of particle i,

$$\mathbf{F}_i = -\frac{\partial V}{\partial \mathbf{r}_i} \tag{4.17}$$

which is one reason why it is convenient that potential functions be written in an explicitly differentiable form.

Let us examine the motion of a particular atom along a particular coordinate, say x. If we know the position at time t, $x(t)$, then the position after a short time interval, Δt, is given by a standard Taylor series:

$$x(t+\Delta t) = x(t)+\dot{x}(t)\Delta t+\ddot{x}(t)\Delta t^2/2+\ldots \tag{4.18}$$

where the dots indicate derivatives with respect to time. The numerical solution to the equations of motion consists of knowing the position $x(t)$, velocity $\dot{x}(t)$, and acceleration $\ddot{x}(t)$, and making suitable approximations to account for the contributions arising from higher derivatives, so that $x(t+\Delta t)$ can be calculated with reasonable precision. Calculating this new position at the end of the time step completes one cycle of the procedure shown schematically in figure 4.1, and the procedure can be repeated iteratively to produce the total trajectory. The acceleration is, of course, given by Newton's second law (4.16), which is, for the one dimensional case,

$$\ddot{x} = F_x/m$$

where m is the atomic mass and F_x is the x component of the total force acting on the atom as a consequence of the combined effects of all the other atoms.

Although (4.18) is exact, the right hand side of the equation is an infinite series that must be truncated in order for $x(t+\Delta t)$ to be determined. One of the fundamental differences between different molecular dynamics algorithms is how they handle this truncation. Fortunately, there is a very simple test to determine how well any algorithm performs. By Newton's third law, the net force acting on the total system of particles is zero, so the total energy of the system (potential plus kinetic) as well as the momentum should be conserved. Small fluctuations or slow drifts in the energy may occur as a consequence of numerical rounding errors, so energy conservation is an important test of an algorithm's performance.

The simplest approach would be to truncate (4.18) after the second derivative, producing the equation for motion with constant acceleration. This is generally a very poor approximation, however, because it attempts to project the motion forward in time based entirely on position, velocity, and acceleration at the beginning of the interval. A variety of improvements to this approximation are possible and appendix 1 gives a summary of the methods that are most commonly used for integrating the equations of motion. These include the algorithms due to Verlet (1967), Beeman (1976) and Gear (1971), as well as the constrained Verlet or SHAKE method (Ryckaert, Ciccotti & Berendsen, 1977; van Gunsteren & Berendsen, 1977).

The molecular dynamics approach was originally developed for simulating atomic motion in simple liquids. The first system to be treated was that of elastic hard spheres (Alder & Wainwright, 1960), followed by models of liquid argon whose atoms interact through a Lennard–Jones potential (Rahman, 1964, 1966; Verlet, 1967). The first polyatomic molecular liquid to be examined was water (Rahman & Stillinger, 1971; Stillinger & Rahman, 1972, 1974). As computers grew larger and faster, it became feasible to use molecular dynamics to simulate macromolecular systems, starting with bovine pancreatic trypsin inhibitor (BPTI), a small globular protein (McCammon, 1976; McCammon, Gelin & Karplus, 1977; van Gunsteren & Berendsen, 1977).

A variety of experimental conditions can be simulated by molecular dynamics. The first simulations on macromolecular systems (McCammon, 1976; McCammon *et al.*, 1977; van Gunsteren & Berendsen, 1977) treated the isolated molecule without any explicit solvent surroundings. These simulations were later modified to include water and neighboring protein molecules, as would be found in a crystal (van Gunsteren *et al.*, 1983). To make the model system, which contains at most a few thousand atoms,

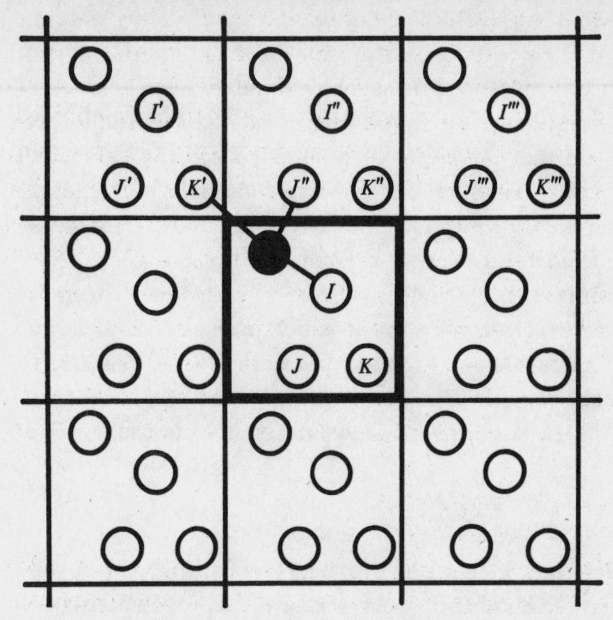

Fig. 4.4. Periodic boundary conditions. A two-dimensional problem is used as an example, with repeating square boundary conditions. The sample system contains four particles in a square (heavy line). It is surrounded by a periodic array of images of the original system. Only one interaction is considered between a particular particle (solid circle) and each of the other three particles in the system (I, J, and K). In each case, the interaction is taken to be that which has the shortest interaction distance. Here, one interaction is with a real particle (I), while two are with images (J'' and K').

represent part of a system of essentially infinite extent, it is conventional to create copies of the model that repeat periodically in every direction. In the most common case, using a cubic geometry with periodic boundary conditions, the sample volume will be a cube embedded in a cubic lattice, each element of which is simply an image of the sample system (figure 4.4). Other geometries for periodic boundary conditions are possible, such as the truncated octahedron, which substantially reduces the volume of the sample (Adams, 1979). Periodic boundary conditions guarantee that all atoms are surrounded by neighbors, either 'real' atoms (those of the original sample), or, if they are beyond the boundary of the sample, images of those atoms. Note that when image atoms are present, the pairwise interaction between atom I and atom J is only to be calculated for the real atom J or whichever of its images is nearest to atom I. This minimum image convention guarantees that multiple interactions between atoms are not included (see figure 4.4).

As discussed above, the total energy of the system being simulated will be conserved if the numerical integration of the equations of motion is accurate enough. This is true for *in vacuo* simulations of macromolecules as well as simulations including solvent and fixed periodic boundary conditions. From the standpoint of a statistical mechanical analysis, the system being simulated under the foregoing conditions corresponds to the microcanonical ensemble (McQuarrie, 1976; Friedman, 1985a); that is, the system has a fixed total energy, volume and number of particles. It is frequently desirable to simulate a system at constant temperature and volume (the canonical ensemble) or to simulate a system under conditions of constant temperature and pressure (the isothermal–isobaric ensemble), since these are the conditions under which most experiments are performed. We note that the temperature is calculated from the atomic velocities, with

$$3k_B T = \sum_{i=1}^{N} m_i \mathbf{v}_i \cdot \mathbf{v}_i / N$$

where k_B is Boltzmann's constant, m_i and $\mathbf{v}_i$ are the mass and velocity of atom i, and N is the number of atoms. For a simulation at constant energy, the temperature fluctuates due to the spontaneous interconversion of the kinetic and potential components of the total energy. To hold the temperature constant, the atomic velocities can be rescaled or otherwise adjusted. To hold the pressure constant, the volume must be allowed to fluctuate by rescaling the interatomic distances and the repeat distances for the periodic boundary conditions. Procedures for properly scaling these quantities to maintain constant temperature and/or pressure have been developed by Andersen (1980) and by Berendsen *et al.* (1984), among others. These methods are summarized in appendix 3.

There are many problems in which it would be desirable to examine motions in a particular region (for example, the binding site in an investigation of interactions between a macromolecule and a ligand) without investing the resources necessary for simulating the entire macromolecule and its solvent surroundings. A study such as this can be carried out by defining a volume centered on the region of interest, within which the normal molecular dynamics simulation is performed. This volume is surrounded by a bath region, a shell within which atoms move according to stochastic dynamics determined, e.g., by the Langevin equation (3.3). The bath region is surrounded, in turn, by a reservoir region, a larger shell in which the atoms have no motion or are replaced by constraining potentials that act on the primary and bath atoms. The outermost reservoir serves as a boundary that preserves the global density and overall structure, while the bath region accommodates the fluctuations in density, structure and energy that occur in the region where full molecular dynamics is carried out. This procedure, molecular dynamics with stochastic boundary conditions, was first tested on a model of liquid argon (Berkowitz & McCammon, 1982), where it was found to give results for the radial distribution function and the velocity correlation function that were in very good agreement with the results from a full molecular dynamics simulation on the same system. This procedure holds considerable promise for those situations where computer time is too limited to allow simulation of the full system and for situations where periodic boundary conditions can be expected to produce artifacts. It has recently been extended to the simulation of proteins, including the active site region of the enzyme, ribonuclease (Brooks, Brünger & Karplus, 1985; Brünger, Brooks & Karplus, 1985). It is known that the presence of confining boundaries can alter the structure of systems such as water (Lee *et al.*, 1984; Belch & Berkowitz, 1985), and that small system sizes can alter density fluctuations (Guillot & Guissani, 1985). More work is needed to develop stochastic boundary methods in which such effects are properly controlled.

Molecular dynamics at low temperatures has been shown to be a useful method for energy minimization, of comparable efficiency to other methods (Brooks *et al.*, 1983). It has been used to anneal local steric conflicts arising during large scale deformation of macromolecular models (Levitt, 1983*c*; Prabhakaran & Harvey, 1985). Molecular dynamics is also a promising method for the refinement of structures from X-ray crystallography (Westhof, Gallion, Weiner & Levy, 1986).

Given the wide range of problems that can be attacked by molecular dynamics, and considering the growing popularity of the method, it is worthwhile to examine some of the limitations of this technique. Most of these grow out of the fact that the molecular models are so detailed (typically including at least all heavy atoms and frequently many or all hydrogens) and the time step is necessarily so short (a fraction of the period of the highest frequency motion, and hence on the order of 1 fs) that only relatively small systems and short time scales are accessible to simulations. Total molecular weights (including solvent) are typically on the order of 25000 or less, and the total period covered is typically a fraction of a nanosecond. With the availability of supercomputers, array processors, and multiprocessors, these values may be pushed up by one or more orders of magnitude, but the accessible time scales will still be far shorter than those of most biologically interesting processes. It is not convenient to use a model that evolves freely under molecular dynamics to examine motions or infrequent processes whose characteristic times are longer than 100 ps or so, because these are unlikely to occur spontaneously during such a short simulation. In some cases, this problem can be overcome by choosing a suitable reaction coordinate and forcing the system to move along that coordinate during a molecular dynamics simulation. These procedures are discussed in more detail in section 4.9.

Another shortcoming of molecular dynamics is that the atomic trajectories are based on classical equations of motion, so that quantum dynamical effects are ignored. Fortunately, this is usually not a serious limitation, and quantum dynamical simulation methods can be used in special cases (cf. section 4.1). Of course, quantum effects are implicitly included in the potential energy function which, as discussed in section 4.2, is designed to model the Born–Oppenheimer surface for the system. Since many of the structural changes accompanying reactions such as enzyme catalysis occur on the time scale of molecular dynamics, it would be possible to gain valuable information on the details of these reactions if molecular dynamics simulations can be coupled to *ab initio* calculations on suitable pieces of the enzyme–substrate complex. Important progress in this direction is being made (Chandrasekhar, Smith & Jorgensen, 1985; Weiner, Singh & Kollman, 1985*b*; Jorgensen, Chandrasekhar, Buckner & Madura, 1986; Madura & Jorgensen, 1986).

4.8 Free energy calculations

Simulation methods that sample the possible atomic coordinate values or 'configurations' of a system of atoms can be used to calculate thermodynamic properties of the system. For example, the entropy of a system is directly related to the number of different configurations that are thermally accessible to it (McQuarrie, 1976; Friedman, 1985*a*). Perhaps the most important of the thermodynamic properties are free energies, which provide the fundamental measures of the stability of a system. In principle, the configurational sampling necessary to determine thermodynamic quantities can be performed by normal mode analysis (section 4.6), molecular dynamics (section 4.7), or by Monte Carlo methods. In practice, conventional Monte Carlo methods are rather inefficient in exploring the configuration space of very large molecules when compared to molecular dynamics (Northrup & McCammon, 1980*a*). Normal mode approaches have other limitations as described below, so that molecular dynamics appears to be the optimal sampling technique presently available.

In what follows, it will be useful to describe free energies as functions of two different sets of independent variables or constraints (Pippard, 1964). The first set, denoted by τ, is the usual set of variables used to define the thermodynamic state of a system (e.g., temperature T, volume v, number of particles N, pressure p, etc.). The particular free energy function that characterizes the stability of a system depends on which thermodynamic constraints are used. The Helmholtz and Gibbs free energies are appropriate for constant N, T, v, and constant N, T, p, respectively. The second set of variables, denoted by ξ, is used to distinguish other features of the system, such as different regions of configuration space. The variables ξ may be discrete or continuous. As a simple example, consider a system comprising a receptor protein, a ligand molecule, and a large number of solvent molecules. Then one might use $\xi = 0$ to label all configurations in which the ligand and receptor are not bound, and $\xi = 1$ to label the bound configurations. In other cases it is helpful to use a continuous ξ, for example to label configurations corresponding to different values of a reaction coordinate. When ξ is continuous and represents particle coordinates, it is common to refer to the free energy $W(\xi)$ as a potential of mean force. If the potential of mean force depends on a single variable ξ, the connection between $W(\xi)$ and the stability of the system can be stated explicitly through the equation

$$W(\xi) = -k_B T \ln \rho(\xi) \tag{4.19}$$

where $\rho(\xi)d\xi$ is the relative probability that the fluctuating system will be found in the interval ξ to $\xi + d\xi$. One normalization convention used later (in section 6.2) is that $\rho(\xi^*) = 1$ for the most probable value ξ^* of ξ; correspondingly, the potential of mean force is measured relative to the value $W(\xi^*) = 0$. Not only can potentials of mean force be determined from detailed dynamical models, as described below; they can also be used to determine the effective forces acting along ξ in simplified dynamical models (cf. sections 4.5, 4.9, 4.10).

Free energies and other thermodynamic functions can be extracted from computer simulations by a variety of methods (McQuarrie, 1976; Bennett, 1976; Valleau & Torrie, 1977; Berendsen, 1985*b*; Beveridge & Mezei, 1985; Friedman, 1985*a*). Entropies, internal energies, and free energies can be easily calculated from the results of normal mode analyses (McQuarrie, 1976) but, of course, share the limitations intrinsic to such analyses. These limitations include the neglect of anharmonic contributions and the neglect of any solvation effects that depend on explicit consideration of the solvent molecules. Improved estimates of macromolecular entropy can be obtained by using the normal mode framework to analyze data from molecular dynamics simulations (Karplus & Kushick, 1981; DiNola, Berendsen & Edholm, 1984; Edholm & Berendsen, 1984; Levy *et al.*, 1984*a*). Here one starts with the mean square position fluctuations of the atoms observed in the full simulation, and constructs a simpler quasiharmonic model with harmonic potentials chosen to produce similar fluctuations. The normal modes and corresponding thermodynamic properties of the quasiharmonic model are then determined. This approach partly corrects for the anharmonicity of the true molecular system, but still neglects contributions due to most multiple minima on the potential energy surface or to detailed solvation effects.

Two other approaches have recently been used to evaluate free energies and other thermodynamic properties of biomolecules. These approaches are relatively demanding of computational resources, but can in principle provide very accurate results. In each of these methods, the potential energy function is modified in the course of a set of simulations. The modifications are designed to move the system from one state to another, corresponding to different values of the variable ξ introduced at the beginning of this section. Data collected during these simulations are used to calculate the free energy and other thermodynamic quantities as functions of ξ.

The first approach has been called umbrella sampling. Although this approach has been used for a variety of applications in studies of liquids and solids (Pangali, Rao & Berne, 1979; Ravishanker, Mezei & Beveridge, 1982; Jorgensen, 1983; Berkowitz, Karim, McCammon & Rossky, 1984; Chandrasekhar *et al.*, 1985), it has been applied in biomolecular studies primarily to determine the relative free energies of different conformational states of biopolymers (Northrup *et al.*, 1982*a*; Mezei, Mehrotra & Beveridge, 1985). In the latter applications, ξ is chosen to be an appropriate conformational coordinate. In a conventional simulation of the kind described in section 4.7, a limited range of ξ will be explored because only configurations having $W(\xi)$ within a few $k_B T$ of the local minimum will be sampled. In umbrella sampling, an expanded range of ξ is explored by adding auxiliary terms to the potential function to shift the local minimum in a desired direction. Simulations are then carried out using the modified potential, and the thermodynamic results are normalized to remove the effects of the auxiliary potentials. The umbrella sampling approach has also been used recently to study the motion of an oxygen molecule inside the protein, myoglobin; in this case ξ is essentially a ligand displacement coordinate (Case & McCammon, 1986).

The equation for umbrella sampling along a coordinate ξ is based on the following definition of the probability density introduced in (4.19):

$$\rho(\xi) = \frac{\int \mathrm{d}\Gamma' \, \delta(\xi' - \xi) \exp\{-\beta V(\Gamma')\}}{\int \mathrm{d}\Gamma' \exp\{-\beta V(\Gamma')\}} \tag{4.20}$$

Here, Γ' represents a particular set of the coordinates of all the atoms in the system (i.e., a point in configuration space), while ξ' and $V(\Gamma')$ are respectively the values of the coordinate ξ and of the potential energy for this configuration. The delta function $\delta(\xi'-\xi)$ is equal to zero for $\xi' \neq \xi$, so that the probability in (4.20) can be interpreted as the sum of Boltzmann probabilities $\exp\{-\beta V(\Gamma')\}/\int d\Gamma' \exp\{-\beta V(\Gamma')\}$ for just those configurations with $\xi' = \xi$. The equation (4.20) is from canonical ensemble (constant N, T, v) theory, so that the corresponding potential of mean force given by (4.19) is a Helmholtz free energy. If one now replaces $V(\Gamma')$ in (4.20) by the modified potential function $V(\Gamma')+U(\xi')$, where $U(\xi')$ is the auxiliary (umbrella) potential that moves the system in the desired direction, the resulting equation can be rearranged to express the corresponding modified density $\rho^*(\xi)$ in terms of the original density,

$$\rho^*(\xi) = C \exp\{-\beta U(\xi)\}\rho(\xi) \tag{4.21}$$

Thus, if a simulation is carried out using the modified potential, a probability density $\rho^*(\xi)$ will be obtained that differs from $\rho(\xi)$; in particular, $U(\xi)$ can be chosen to yield a $\rho^*(\xi)$ that is large and statistically reliable in regions of ξ where the density $\rho(\xi)$ obtained from a simulation of the 'real' system would be small and uncertain. Since $U(\xi)$ is a known function, (4.21) can be used to determine $\rho(\xi)$ to within a constant factor in such regions.

The actual procedure for umbrella sampling typically makes use of the potential of mean force rather than the equivalent probability density. One might start by performing a simulation with the unmodified potential V; from the observed configurations, $\rho(\xi)$ and $W(\xi)$ can be determined reliably within a certain range of ξ. An umbrella term U is then added to the potential and another simulation is performed to determine $\rho^*(\xi)$ in a range that partly overlaps the previous range. By using (4.19) and (4.21), the potential of mean force in this new range is, within a constant term, $-k_B T \ln \rho^*(\xi) - U(\xi)$. The constant term is chosen to ensure that the potentials of mean force agree in the region of overlap in the two simulations. By using new umbrella potentials that increasingly shift the sample density along ξ in successive simulations, this procedure may be repeated to obtain the free energy $W(\xi)$ over a considerable range. Other thermodynamic quantities, such as the internal energy and entropy of the system as functions of ξ, can also be determined from these simulations (Northrup *et al.*, 1982*a*).

Another approach for calculating free energies that has a long history of application in liquid state theory is statistical mechanical perturbation theory (McQuarrie, 1976; Postma, Berendsen & Haak, 1982; Friedman, 1985*a*). In the canonical ensemble framework, the fundamental equation for this approach is derived from the definition of the Helmholtz free energy

$$A = -k_{\mathrm{B}}T\left(\ln\frac{\int \mathrm{d}\Gamma \exp\{-\beta V(\Gamma)\}}{N!\,\Lambda^{3N}}\right) \tag{4.22}$$

Here, N is the number of particles in the system, Λ is a function of the temperature, and the other quantities are defined as in (4.20). Now suppose that one wishes to evaluate the difference between the free energy A^* of a system whose potential energy function is $V^*(\Gamma)$ and the free energy A of the system specified above. Using (4.22) to express the two free energies, one easily obtains

$$\begin{aligned} A^* - A &= -k_{\mathrm{B}}T\ln\left[\frac{\int \mathrm{d}\Gamma \exp\{-\beta V^*(\Gamma)\}}{\int \mathrm{d}\Gamma \exp\{-\beta V(\Gamma)\}}\right] \\ &= -k_{\mathrm{B}}T\ln\left[\frac{\int \mathrm{d}\Gamma \exp\{-\beta(V^* - V)\}\exp(-\beta V)}{\int \mathrm{d}\Gamma \exp(-\beta V)}\right] \\ &= -k_{\mathrm{B}}T\ln\langle\exp\{-\beta(V^* - V)\}\rangle_V \end{aligned} \tag{4.23}$$

In the last line of (4.23), the symbol $\langle\rangle_V$ indicates an average over a Boltzmann sample of configurations governed by the potential energy function $V(\Gamma)$. If a simulation is carried out based on $V(\Gamma)$, this average can be determined simply by evaluating the function $\exp\{-\beta(V^* - V)\}$ for the resulting configurations and dividing the sum of these values by the number of configurations. Although the derivation indicated above is for the canonical ensemble, the last line of (4.23) may be used to calculate differences of Gibbs free energies (replace A and A^* by G and G^*) if the reference simulation has constant N, T, p instead of constant N, T, v.

In a typical application, one might compute the difference in free energies of two systems, one of which comprises the solute molecule L plus solvent, while the other comprises a different solute molecule M in the same solvent. Suppose for simplicity that L and M contain the same number of atoms. Denote the potential energy functions for the two systems (solute plus solvent) by V_{L} and V_{M}, respectively. Define a 'mixed' potential energy function

$$V_\lambda = \lambda V_{\mathrm{M}} + (1-\lambda)V_{\mathrm{L}} \tag{4.24}$$

where λ may vary from 0 to 1. If the free energy of the system governed by V_λ is now denoted by $A(\lambda)$, then the free energies of the L/solvent and M/solvent systems are $A(0)$ and $A(1)$, respectively. To calculate the desired difference $A(1) - A(0)$, one could begin by performing a simulation of the

L/solvent system. The configurations generated in this simulation could then be used to compute $A(\lambda)-A(0)$ through (4.23). That is, one evaluates for each configuration the change in energy produced by replacing the L/solvent potential function by the mixed potential function, and then performs the average indicated by (4.23). In practice, statistically reliable values will only be obtained for a range of λ corresponding to free energy differences up to perhaps 5 or 10 times $k_B T$. Outside of this range, relatively few configurations generated with Boltzmann probabilities proportional to $\exp(-\beta V_L)$ will occur and contribute to the average in (4.23). If the range of reliability is too small to determine $A(1)$–$A(0)$, one can obtain this quantity by making use of bridging simulations. For example, if $A(\lambda)$–$A(0)$ is well determined up to $\lambda = 0.5$, one can carry out a simulation governed by the mixed potential (4.24) for $\lambda = 0.5$ and obtain the desired difference as the sum of $A(1)$–$A(0.5)$ from the second simulation and $A(0.5)$–$A(0)$ from the first. In the case of large differences in free energy, it may be necessary to make use of several bridging simulations. The method can also be used to extrapolate from larger to smaller values of λ. In the above example, one could obtain $A(1)$–$A(0)$ by extrapolating in both directions from the simulation performed with $\lambda = 0.5$.

The perturbation method has been used to determine the free energy required to create spherical cavities in water (Postma *et al.*, 1982) and the relative free energies of hydration of small molecules and ions (Jorgensen & Ravimohan, 1985; Lybrand, Ghosh & McCammon, 1985; McCammon, Lybrand, Allison & Northrup, 1986*c*). For many applications, however, it is better to use an extended approach called the thermodynamic cycle–perturbation method (Tembe & McCammon, 1984; McCammon *et al.*, 1986*c*; Lybrand, McCammon & Wipff, 1986; McCammon, Karim, Lybrand & Wong, 1986*a*; Wong & McCammon, 1986*a*, *b*). This approach can be illustrated by considering the binding of two different ligands, L and M, to a receptor molecule R. The relative binding affinity is determined by the free energy difference $\Delta\Delta A = \Delta A_2 - \Delta A_1$ for the following processes

$$\mathrm{L+R \rightarrow LR} \quad \Delta A_1 \tag{4.25}$$

$$\mathrm{M+R \rightarrow MR} \quad \Delta A_2 \tag{4.26}$$

In principle, ΔA_1 and ΔA_2 could be determined by umbrella sampling simulations in which the ligands and the receptor are gradually brought together from a large separation. In practice, this approach will often be frustrated by slow desolvation processes (especially in the case of recessed binding sites, where the entering ligand may hinder solvent escape) or conformational changes in ligand or receptor associated with binding. Even where such complications are not present, each ΔA determination would require a sequence of long simulations, corresponding to the displacement of the ligand relative to the receptor. In the thermodynamic cycle—perturbation approach, one would instead consider the hypothetical reactions

$$L + R \rightarrow M + R \quad \Delta A_3 \tag{4.27}$$

$$LR \rightarrow MR \quad \Delta A_4 \tag{4.28}$$

Since reactions (4.25) through (4.28) form a closed thermodynamic cycle, $\Delta\Delta A = \Delta A_2 - \Delta A_1 = \Delta A_4 - \Delta A_3$. The free energy changes ΔA_3 and ΔA_4 can be calculated by the perturbation method described above. The thermodynamic cycle—perturbation approach has a number of important advantages. First, it provides accurate results because detailed solvation effects and molecular anharmonicity are properly accounted for, at least within the framework of classical dynamics. Second, the computational requirements for determining ΔA_3 and ΔA_4 are typically much smaller than those for calculating ΔA_1 and ΔA_2. Third, by focusing on relative free energy changes, the approach benefits from cancellation of terms that may be difficult to evaluate. Such terms include some of those associated with nonideality of the solutions (e.g., the similar contributions to ΔA_1 and ΔA_2 due to charge screening if L and M carry the same charge), with the internal dynamics of the ligands (e.g., the similar contributions to ΔA_3 and ΔA_4 associated with alterations of restoring forces or with changes in the number of atoms), and with quantum dynamical effects (e.g., the similar contributions to ΔA_3 and ΔA_4 due to the classical treatment of high frequency vibrations).

The thermodynamic cycle—perturbation approach has been used successfully to calculate the relative affinity of different halide anions for a synthetic ionophore (McCammon *et al.*, 1986*c*; Lybrand *et al.*, 1986); different inhibitors for the enzyme, trypsin (McCammon *et al.*, 1986*a*; Wong & McCammon, 1986*a*, *b*); and a given inhibitor for site-specific mutants of trypsin (Wong & McCammon, 1986*a*, *b*). It should also be useful in calculations of relative free energies of assembly for different biomolecular structures, estimation of relative rate constants by calculation of relative free energies of activation, and other problems. The approach is new enough that substantial refinement and modification of the specific procedures described above can be anticipated. For example, it may be advantageous in some cases to use a method called thermodynamic integration instead of perturbation theory to compute free energy changes such as those indicated in (4.27) and (4.28). Here, one would use the basic equation

$$A(\lambda = 1) - A(\lambda = 0) = \int_0^1 \frac{\partial A(\lambda)}{\partial \lambda} \mathrm{d}\lambda$$

$$= \int_0^1 \left\langle \frac{\partial V_\lambda}{\partial \lambda} \right\rangle_\lambda \mathrm{d}\lambda$$

in place of (4.23) (Bennett, 1976; Berendsen, 1985*b*; Beveridge & Mezei, 1985). As in the case of (4.23), the free energy obtained is a Helmholtz free energy if N, T, v are constant and a Gibbs free energy if N, T, p are constant. If V_λ is of the form (4.24), one obtains

$$A(\lambda = 1) - A(\lambda = 0) = \int_0^1 \langle V_M - V_L \rangle_\lambda \mathrm{d}\lambda$$

This method may require less computer time than the perturbation method if the change in free energy is quite large (D. Beveridge, M. Mezei, W. van Gunsteren, personal communications). Also, it may be advantageous to replace (4.24) by a function that is nonlinear in λ in some situations (A. Cross, M. Mezei, personal communications).

4.9 Activated molecular dynamics

Activated processes are of widespread importance in biology. Examples of such processes range from local conformational changes associated with the binding of ligands (Debrunner & Frauenfelder, 1982) to the rearrangement of covalent bonds that occurs in enzymatic reactions (Lipscomb, 1982). To determine the dynamical details of activated processes, it is desirable to carry out simulations of the atomic motions involved. However, as described in section 4.1, the conventional molecular dynamics technique is not sufficient for such simulations due to the infrequent occurrence of the transitions.

Recently, a modified molecular dynamics method has been developed for simulation studies of such processes (McCammon & Karplus 1979, 1980*b*; Rosenberg, Berne & Chandler, 1980; Northrup *et al.*, 1982*a*). In this method, the calculation is divided into two parts. In the first part, one calculates the probability of finding the system in a region near the top of the free energy barrier. In the second part, trajectory calculations are initiated from the region of the barrier top, thereby avoiding the long activation step. The approach is most easily understood in terms of the following expression for the rate constant (Northrup *et al.*, 1982*a*)

$$k = (\tfrac{1}{2})\kappa\langle|\dot{\xi}|\rangle\left[\rho(\xi^\ddagger)\Big/\int_i \rho(\xi)\mathrm{d}\xi\right] \qquad (4.29)$$

Here, κ is a transmission coefficient (see below), ξ is a reaction coordinate that measures the progress of a transition, $\dot{\xi}$ is the time derivative of ξ,

$\xi^\ddagger$ designates the value of ξ at the peak of the barrier (the transition state), and $\rho(\xi)d\xi$ is the probability of finding the system in the interval ξ to $\xi + d\xi$. Except for κ, all of the factors that appear in this expression are time-average or equilibrium quantities. The transmission coefficient κ depends on the detailed dynamics of the trajectories passing through the barrier region.

The first step of the modified technique consists of defining a suitable reaction coordinate for the process and calculation of $\rho(\xi)$; this provides a value for the factor in square brackets in (4.29). The reaction coordinate is a function of the system coordinates and has the following ideal properties: (a) the reaction coordinate varies monotonically as the system moves from the initial state region over the energy barrier and into the final state region, (b) all configurations with ξ less than (greater than) the barrier top value $\xi^\ddagger$ experience a mean force, averaged with respect to coordinates other than ξ, in the direction of the initial (final) state region. From property (b), it is desirable that ξ include terms corresponding to all interactions that make sizable and systematic contributions to the barrier.

Given the reaction coordinate, the calculation of $\rho(\xi)$ can be carried out by an umbrella sampling procedure, as described in section 4.8 (Northrup *et al.*, 1982*a*). In this procedure, one carries out a sequence of simulations in which the system is constrained to remain within small intervals of ξ, but with unrestricted sampling of all other coordinates. The mean temperature should be the same in each simulation. By piecing together data from overlapping 'windows' of ξ, $\rho(\xi)$ can be constructed over a range extending from the reactant region through the barrier region. In the second part of the calculation, a large number of independent trajectories is calculated by the molecular dynamics technique, starting from a representative set of states with $\xi = \xi^\ddagger$. A convenient procedure for generating these initial states is to carry out a dynamical simulation with the umbrella term $U(\xi) = (k/2)\,(\xi - \xi^\ddagger)^2$ added to the potential energy function. If k is sufficiently large, this simulation will yield many states (sets of atomic coordinates and velocities) in the region of the barrier peak. From these states, one selects a smaller number to be used as initial conditions for trajectories propagated in the absence of the umbrella potential. These states should have $\xi = \xi^\ddagger$ to avoid artifacts associated with the removal of the umbrella potential. Also, they should be sufficiently separated in time to be dynamically uncorrelated, so that successive trajectories will yield new information. Finally, enough initial states should be chosen to provide a representative sample of the different configurations that may lie within the barrier region. In practice, the last two requirements may necessitate the generation of several hundred or a thousand initial states with a minimum time separation of 0.3 ps (Northrup *et al.*, 1982*a*). The subsequent trajectories are propagated until the system settles into a stable state; the average time for this may be about 0.5 ps or longer, depending on the system.

The trajectory data from the simulations described above are used to calculate $\langle|\dot{\xi}|\rangle$ and to evaluate κ according to the following equation (Chandler, 1978)

$$\kappa(t) = D\langle\dot{\xi}(0)\delta[\xi(0)-\xi^{\ddagger}]H_p[\xi(t)]\rangle \tag{4.30}$$

Here, D is a normalization constant chosen such that $\kappa(0_+) = 1$, δ is the Dirac delta function, and H_p is a step function which is equal to 1 or 0 for ξ greater than or less than $\xi^{\ddagger}$, respectively. Thus $\kappa(t)$ is a time correlation function that measures the net flux into the final state region for trajectories that start at the barrier top. This reactive flux time correlation function decays rapidly (usually within a picosecond) to a plateau value that is used in equation (4.29). The trajectories can also be propagated forward and backward in time to generate a representative set of barrier crossing events. Analysis of such trajectories provides a wealth of detailed information concerning the mechanism of the activated process (McCammon, Lee & Northrup, 1983).

4.10 Brownian dynamics

Motions of proteins or nucleic acids that involve substantial displacements of the molecular surface are heavily damped due to the high viscosity of the surrounding solvent. In many cases, the damping effects are sufficiently large that inertial forces are negligible and the motion has a Brownian or random walk character; the dominant forces are then of the kinds that appear on the right-hand side of equation (3.3). Detailed simulations of such motions can be carried out by the Brownian dynamics method, which is a diffusional analog of the molecular dynamics method (Ermak & McCammon, 1978). As with molecular dynamics, a number of different algorithms have been developed to calculate diffusional trajectories; recent examples and additional references can be found in papers by Helfand (1984), Lamm (1984), Dickinson (1985) and Fixman (1986). The method has been used to study both intramolecular motions, e.g., those of polypeptides and short DNA double helices, and intermolecular motions, e.g., the diffusional encounter of enzyme and substrate. Procedures analogous to those described in section 4.9 can be used to study diffusional barrier-crossing processes (Northrup & McCammon, 1980*b*).

The Brownian dynamics method is perhaps most easily illustrated by considering its application to a single spherical molecule moving along the x axis in the absence of any applied force. Suppose that the molecule is initially at the point x_0. Then, from the analytic solution of the diffusion equation, the probability density for finding the molecule at a point x after a time Δt is just the Gaussian function

$$\rho(x, \Delta t) = (4\pi D\Delta t)^{-\frac{1}{2}} \exp[-(x-x_0)^2/4D\Delta t] \tag{4.31}$$

The first step of the simulation is completed by choosing the position of the molecule at random from this probability distribution. In other words, the algorithm for the first step of a diffusional trajectory is

$$x = x_0 + R$$

where R is a random number with the statistical properties

$$\langle R \rangle = 0$$

$$\langle R^2 \rangle = 2D\Delta t$$

The trajectory is extended to times $t = 2\Delta t$, $t = 3\Delta t$, etc., by repeated application of the above algorithm, taking the position at the start of each step to be that chosen in the preceding step. The procedure yields one representative diffusional trajectory. By computing a large number of such trajectories with different sets of random numbers, one generates a description of how an ensemble of diffusing molecules behaves. For instance, as the number of trajectories grows larger and larger, the ensemble position distribution at time Δt approaches the result given in equation (4.31).

The above method has been extended to systems that are not amenable to exact analytic analysis. For example, for the relative motion of a pair of spherical reactant molecules that interact by a noncentrosymmetric force $\mathbf{F}(\mathbf{r})$ but not by hydrodynamic interactions, one has the algorithm

$$\mathbf{r} = \mathbf{r}_0 + \beta \mathrm{D}\mathbf{F}(\mathbf{r}_0)\Delta t + \mathbf{R} \tag{4.32}$$

where $\beta = (k_B T)^{-1}$ and

$$\langle \mathbf{R} \rangle = 0$$

$$\langle R_i R_j \rangle = 2D\delta_{ij}\Delta t$$

D is the relative diffusion coefficient, and i, j label the Cartesian components of $\mathbf{R}$. In generating trajectories with equation (4.32), the time step Δt must be short enough that $F(\mathbf{r}) \approx F(\mathbf{r}_0)$. Similar algorithms are available for the cases of many spherical particles with arbitrary forces and hydrodynamic interactions (Ermak & McCammon, 1978), nonspherical particles modeled as flexible or rigid assemblies of spherical subunits (Allison & McCammon, 1984*a*, *b*), and spheres with translational and rotational motion explicitly coupled by hydrodynamic interaction (Dickinson, Allison & McCammon, 1985).

The diffusional motions of single molecules can be analyzed by Brownian dynamics, as described above, or by analytic modeling in certain special cases. For ellipsoidal particles, analytical expressions are known for the friction and diffusion coefficients (Perrin, 1934; Cantor & Schimmel, 1980), and simulations of the Brownian motion of these particles using equation (4.31) is straightforward (Harvey & Cheung, 1972). Particles with more complex shapes are generally modeled as assemblies of spherical subunits, often called 'beads' (Kirkwood & Riseman, 1948; Bloomfield, Dalton & Van Holde, 1967; McCammon, Deutch & Felderhof, 1975; McCammon & Deutch, 1976; Garcia de la Torre & Bloomfield, 1981; Allison & McCammon, 1984a). Formalisms have been developed to treat flexible molecules consisting of rigid subunits connected by hinges or swivels (Harvey, 1979; Wegener, Dowben & Koester, 1980), and the bead methods have been extended to treat such models when the subunits have irregular shapes (Wegener, 1982; Harvey, Mellado & Garcia de la Torre, 1983).

To obtain rate constants for biomolecular diffusion controlled reactions by Brownian trajectory calculations, one first divides the diffusion space around one of the reactants (e.g., an enzyme) into two regions. The 'inner' region is finite and comprises that volume adjacent to the enzyme in which the interactions are complicated and best dealt with by numerical techniques. The 'outer' region is of infinite volume but is everywhere far enough from the enzyme that diffusional behavior can be described analytically. Trajectories then need be computed only in the inner region. Any trajectory that reaches the outer region is truncated; the contribution of such trajectories to the rate is given by an analytic correction factor.

The factors needed in the rate constant calculation can be introduced successively as follows (Northrup, Allison & McCammon, 1984; Allison, Northrup & McCammon, 1985*b*; Northrup, Curvin, Allison & McCammon, 1986; Northrup, Smith, Boles & Reynolds, 1986). Let the target molecule (e.g., an enzyme) be surrounded by a spherical surface of radius b, where b is large enough that the surface lies just outside of the inner region. Then the rate constant can be written as

$$k = k_D(b)\beta_\infty \tag{4.33}$$

Here, $k_D(b)$ is the steady state rate at which mobile reactants with $r > b$ would first strike the surface. Since diffusion up to the distance b can be described analytically, $k_D(b)$ will be given by a known expression; e.g., if the enzyme–substrate interactions are negligible for $r > b$, one has the familiar Smoluchowski result

$$k_D(b) = 4\pi Db \tag{4.34}$$

The factor β_∞ is the probability that a reactant starting at $r = b$, and free to diffuse in the inner and outer regions, will react rather than escape. The probability of escape from $r = b$ is then $1-\beta_\infty$. Now let the b surface be enclosed by a larger spherical surface of radius q. The probability of escape from $r = b$ can be written as

$$1-\beta_\infty = (1-\beta)P_\infty \tag{4.35}$$

where $1-\beta$ is the probability that a reactant starting at $r = b$ will reach $r = q$ before reacting, and P_∞ is the probability that a reactant starting at $r = q$ will escape rather than react. Although evaluation of P_∞ formally still requires consideration of reactant motion in the inner and outer regions, it can be shown that the contribution of the inner region dynamics can be expressed in terms of β as (Northrup *et al.*, 1984)

$$P_\infty = (1-\Omega)[1-(1-\beta)\Omega]^{-1} \tag{4.36}$$

where $$\Omega = k_D(b)/k_D(q) \tag{4.37}$$

Combining equations (4.33)–(4.37), one obtains

$$k = k_D(b)\beta[1-(1-\beta)\Omega]^{-1} \tag{4.38}$$

Thus, to calculate the rate constant for a bimolecular diffusion controlled reaction in dilute solution, one need only compute a large number of trajectories of one reactant diffusing in the vicinity of the other, fixed reactant. Trajectories are initiated at $r = b$ and terminated upon reaction or upon reaching $r = q$. The fraction of trajectories that react is β, and this quantity yields the rate constant through equation (4.38). Analysis of the trajectories also provides information on the mechanistic details of the reaction, e.g., whether reactants tend to be steered into productive collision geometries during the diffusional encounter.

5

Short time dynamics

5.1 Introduction

In this chapter, we consider the types of motion that occur in proteins and nucleic acids during time intervals of less than 100 ps. Motions on these time scales are relatively well characterized theoretically because they can be simulated by conventional molecular dynamics techniques. Although most biological activity occurs over much longer periods of time, the types of motion described here are of interest for a variety of reasons. As indicated in chapter 3, fast motions represent a kind of dynamic background that partly determines the nature of all slower motions. For example, the frictional effects that arise from atomic collisions are essentially fully developed within a few picoseconds. Also, many structural transitions that seem to be characterized by long times are in fact fast motions that happen to occur infrequently. It was noted in section 4.9 that activated processes often involve such intrinsically fast transitions. Certain transitions that are triggered by photon absorption or electron transfer may also be intrinsically fast. The mechanistic details of such transitions can be expected to have features similar to those seen in typical fast motions. Of practical interest is that proteins and nucleic acids may explore a fairly representative sample of the conformations in the regions of their native states in time intervals of between 10 and 100 ps. As will be discussed in chapter 9, this means that useful thermodynamic quantities, such as relative free energies of ligand binding, can be obtained from simulations on these time scales. Finally, a variety of experiments provide information on picosecond time scale motions. Molecular dynamics calculations can aid in the interpretation of data from such experiments and these data, in turn, provide useful tests of the theoretical models.

5.2 Results for proteins

Several dozen simulations of protein dynamics have been carried out to date. These have involved a number of proteins, most of which have been the subject of several studies based on different computer models. A consistent picture of the nature of short time scale motions has emerged from this work. In this section, we first describe some of the important structural features of protein motions. We then describe how simple physical models can be used to help analyze these motions, and how the simulation results relate to experimental data. The surroundings of the protein molecules have not been explicitly included in most simulations that have been reported so far. The few exceptions include simulations of the small proteins bovine pancreatic trypsin inhibitor and avian pancreatic polypeptide (BPTI and APP) in hydrated crystals (van Gunsteren *et al.*, 1983; Krüger *et al.*, 1986; Berendsen, van Gunsteren, Zwinderman & Geurtsen, 1986), BPTI in an apolar solvent (van Gunsteren & Karplus, 1982*b*), BPTI and APP in water (van Gunsteren & Berendsen, 1984; Krüger *et al.*, 1985) and the enzyme trypsin in water (Wong & McCammon, 1986*a,b*). In what follows, we focus initially on results obtained without explicit surroundings since these have been analyzed in the greatest detail. We then briefly consider the results obtained with explicit surroundings; in general, these are consistent with the former results except at the protein surface, where there are moderate differences in relaxation times due to solvent damping and in displacement amplitudes of some atoms due to protein–protein contacts in the crystal simulations.

5.2.1 *Local aspects*

Atoms near the surface of a protein generally exhibit larger thermal displacements than do those in the interior of the molecule. This behavior reflects the rough topography of protein surfaces (cf. figure 2.5); exposed segments of the molecule are subject to relatively weak steric constraints. For cytochrome c at 300 K, the average root-mean-square (rms) position fluctuation of the atoms is on the order of 0.05 nm in the protein interior (Northrup *et al.*, 1981). This average remains nearly constant with increasing distance from the protein center until the region of the rough molecular surface is reached. In this exterior region of the protein, the average rms position fluctuation of the atoms grows rapidly to values on the order of 0.1 nm. Similar results have been obtained in molecular dynamics studies of other proteins. For example, the largest motions in the small protein BPTI occur in the loops and ends of the polypeptide chain at the surface of the protein (McCammon *et al.*, 1977; van Gunsteren & Karplus, 1982*b*; Swaminathan, Ichiye, van Gunsteren

& Karplus, 1982; Levitt, 1983*b*). Two other correlations between structure and amplitude have been widely observed. Atoms that belong to elements of secondary structure such as alpha helices or beta sheets tend to exhibit relatively small displacements. In part, this correlation may reflect the tendencies for these structural elements to be buried and for the loops that connect these elements to be near the surface. At a more local level, the position fluctuations are usually larger for atoms in sidechains than for those in the polypeptide backbone. Moreover, within a typical sidechain, the fluctuations become larger as one moves away from the backbone.

The motion of a typical atom in a protein is quite anisotropic (Northrup *et al.*, 1981; Morgan, McCammon & Northrup, 1983). That is, the displacements are significantly larger in some directions than in others. The anisotropy can be described quantitatively in terms of the mean square displacement matrix of each atom. The mean square displacement matrix for a particular atom is (Willis & Pryor, 1975)

$$\begin{pmatrix} \langle \Delta x \Delta x \rangle & \langle \Delta x \Delta y \rangle & \langle \Delta x \Delta z \rangle \\ \langle \Delta y \Delta x \rangle & \langle \Delta y \Delta y \rangle & \langle \Delta y \Delta z \rangle \\ \langle \Delta z \Delta x \rangle & \langle \Delta z \Delta y \rangle & \langle \Delta z \Delta z \rangle \end{pmatrix}$$

where Δx, Δy and Δz are the displacements of the atom from its mean position along the x, y and z axes, respectively, and the angular brackets indicate time averages. The sum of the diagonal elements is equal to the mean square position fluctuation of the atom. For each atom, a rotated set of Cartesian axes can be defined in which the mean square displacement matrix is diagonal; these axes are the principal axes of the matrix. The three nonzero elements of the diagonalized matrix are the matrix eigenvalues. Each eigenvalue is equal to the mean square fluctuation of the atom along one of the principal axes. For cytochrome c, the ratio of the mean square displacements along the axes of smallest and largest displacement is about 0.26 for a typical atom; the range of such ratios observed is 0.03 to 0.68 (Northrup *et al.*, 1981).

For certain classes of atoms, the direction of preferred displacement can be correlated with the local protein structure (Morgan *et al.*, 1983). An example is shown in figure 5.1. The atoms CD1 and CD2 at the end of the sidechain have major axes that are roughly consistent with rotational oscillations about the single bond between the CB and CG atoms. In general, however, the directions of preferred displacement are determined by collective motions that are unrelated to local bonding; this effect is discussed below.

As mentioned before, the atomic motion in proteins is substantially

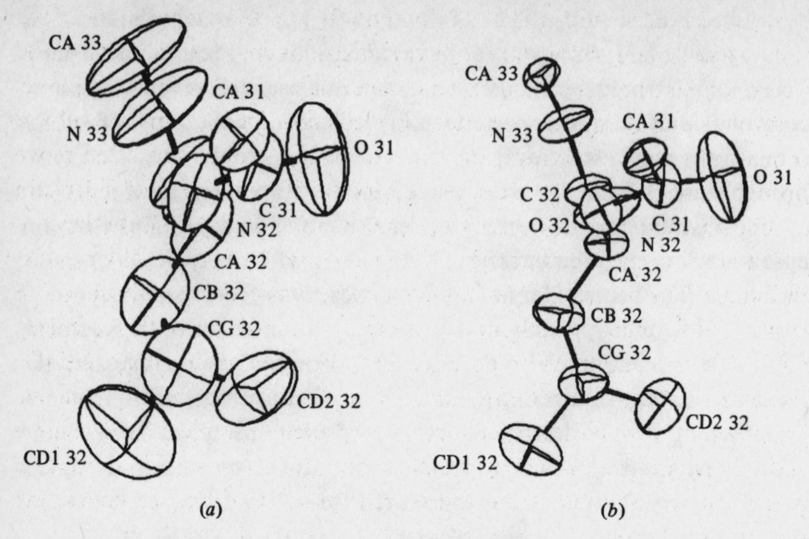

Fig. 5.1. Thermal ellipsoids representing the position fluctuations of the atoms in the leucine 32 residue of cytochrome c, based on a molecular dynamics simulation (Morgan *et al.*, 1983). The lengths of the principal semiaxes correspond to (*a*) the rms fluctuations observed during a period of 32 ps, (*b*) twice the rms fluctuations observed during periods of 0.2 ps. The importance of collective motion during the longer time period is evident in the similar orientations of the longest axes of a number of atoms. (Copyright © 1983, John Wiley & Sons, Inc. Reprinted by permission.)

anharmonic at room temperature. This can be seen directly by calculating moments of the atomic position distribution functions. For harmonic motions, these distribution functions will be Gaussian. In this case, the first few moments of the position distribution functions along a given direction satisfy the following equalities

$$\langle(\Delta x)^3\rangle = 0 \tag{5.1}$$

$$\langle(\Delta x)^4\rangle = 3\langle(\Delta x)^2\rangle^2$$

Calculations based on molecular dynamics simulations of cytochrome c show that one or both of these equalities are substantially violated for at least half of the atoms in the protein (Mao, Pear, McCammon & Northrup, 1982*a*). Moreover, the anharmonic effects are related to anisotropy; for a typical atom, the anharmonic effects are largest for motion in the direction of largest displacement. The atoms with largest anharmonicity fall into two classes. In the first class are those atoms whose effective potentials (potentials of mean force) are reasonably symmetric about the average atomic positions, but the potentials are relatively flat near the center (square well distortion) or have similar minima on either side of the center. Anharmonicity of this type is especially common for atoms near the protein surface and at the ends of sidechains. Such atoms are subject to relatively weak structural constraints; surface groups can often move through some distance before encountering steric repulsion, and sidechain termini typically have enhanced conformational mobility because they are separated from the comparatively rigid backbone by rotationally permissive bonds. The second class of anharmonic atoms comprises those whose effective potentials typically have a dominant minimum with secondary minima on one side; such potentials lead to skewed position distributions. The existence of multiple minima in the potentials of mean force for displacement of many atoms indicates that proteins sample different stable conformations, all of which are similar to the native structure, in the course of their dynamics at normal temperatures. As will be discussed below, the transitions between these conformational substates typically have collective character, and the existence of these substates is significant in the interpretation of certain experimental results.

The importance of anharmonic effects in protein dynamics has also been shown in a study of the shape of the potential surface of BPTI near an energy minimum in the vicinity of the native state (Gō *et al.*, 1983). In this study, a normal mode calculation was first carried out in the dihedral angle space of the protein. The atomic positions were then displaced in accordance with each normal mode; the actual potential energy of the distorted protein was calculated and compared with that obtained by a quadratic fit to the surface near its minimum. For displacements of typical thermal magnitude, significant differences in the two energies were found for the softest directions. Of the 36 modes with frequencies below $\nu = 1\ \mathrm{ps}^{-1}$, only one is close to being harmonic. These low frequency anharmonic modes were found to dominate the atomic displacement magnitudes. Superimposed on these low frequency motions are localized motions with higher frequency and smaller amplitude; these localized motions are more harmonic in character. Subsequent normal mode studies of BPTI (Brooks & Karplus, 1983) and of BPTI, crambin, ribonuclease, and lysozyme (Levitt *et al.*, 1985), have confirmed the general results described above; see section 5.2.2.

Also of interest is a study of the decaglycine alpha helix (Levy, Perahia & Karplus, 1982). Here, the atomic position fluctuations from a molecular dynamics simulation (reflecting motion on the true potential surface) were compared with those from a normal mode calculation. Substantial anharmonicity was apparent even in this comparatively rigid element of secondary structure. At room temperature, the mean square displacements observed in the dynamical simulation are approximately twice as large as those in the corresponding harmonic model. Again, detailed analysis showed that the low frequency motions are primarily responsible for the atomic displacements and display the largest anharmonicities.

The preceding discussion has only considered time-average properties of the atomic displacements. In considering the time dependence of the atomic motions, it is appropriate first to examine the dependence of the apparent amplitude of atomic motion upon the observation period (Morgan *et al.*, 1983). The rms displacements of the protein atoms approach their limiting values in a time on the order of 10 to 20 ps. About one-half of this limiting value is achieved in a time of 1 ps. For the short time (less than 1 ps) position distribution functions, the anisotropy of atomic displacement is simply related to the local bonding. That is, the atomic displacements are dominated by local rotations about single bonds. This result is quite apparent in figure 5.1(*b*). These correlations are largely washed out by collective motions in the protein at longer times, as is apparent in figure 5.1(*a*).

More detailed information on the dynamics of short time motions can be obtained by calculation and analysis of appropriate time correlation functions (McCammon, 1976; McCammon *et al.*, 1977; Swaminathan *et al.*, 1982; Levitt, 1983*b*). The time correlation function $C_A(\tau) = \langle A(t+\tau)A(t)\rangle$ for a dynamical variable A is obtained by multiplying the value of A at the time t by its value at the subsequent time $t+\tau$, calculating such products for a representative set of initial times t, and averaging. If A is the fluctuation of a variable from its mean value, $C_A(0)$ is the mean square fluctuation of the variable for an equilibrated system, while the subsequent values $C_A(\tau)$ describe the average manner in which a fluctuation decays. The time correlation functions for displacements of atom positions or for rotations about single bonds typically indicate the presence of different processes occurring on two or more different time scales. The time correlation functions generally decay significantly in the first 0.2 ps. During the subsequent few ps, a variety of behaviors is observed, ranging from slow, nearly monotonic decay to damped oscillation ($\nu < 1\ \text{ps}^{-1}$).

Simulations of proteins in crystal or solvent environments have confirmed the general conclusions described above while introducing a number of interesting variations. Van Gunsteren *et al.* (1983) carried out a 20 ps simulation of the full unit cell of a BPTI crystal subject to periodic boundary conditions. The cell contained four protein molecules and 560 water molecules. Perhaps the most interesting finding was that the four BPTI molecules rapidly assumed slightly different, quasistable conformations, corresponding to four distinct low energy regions on the potential energy surface. The rms atomic differences between these mean structures (0.16–0.21 nm) were similar to the differences of each from the X-ray structure. This result confirms the conclusion from previous simulations and from experimental studies (section 5.2.4) that proteins, in addition to their rapid vibrations, undergo transitions among different stable substates that are similar to the X-ray structure. The magnitudes of the atomic fluctuations in each molecule, and the structural variations of these magnitudes (e.g., sidechain versus backbone, exposed loops versus buried β sheet), are similar to those found earlier. Where solvent molecules are observed to be coordinated with surface atoms of the protein in the X-ray structure, similar coordination is seen in the simulation. The mobility of the solvent molecules increases with distance from the protein. Qualitatively similar results have been obtained in a 40 ps simulation based on a different crystal form of BPTI (Berendsen *et al.*, 1986).

The first simulation of a protein (BPTI) in an explicit solvent environment made use of a simple apolar solvent constructed to have a density similar to water (van Gunsteren & Karplus, 1982*b*; Swaminathan *et al.*, 1982). Here, the most significant difference from the simulations *in vacuo* was a slowing of the atomic displacements, especially at the protein surface where the correlation times were increased by a factor of about two due to solvent friction. The amplitudes of atomic displacement were not significantly changed. Subsequently, several simulations in explicit aqueous environments have been performed. For BPTI, van Gunsteren & Berendsen (1984) found that the average solution structure is similar to the crystal structure. The primary differences are displacements of several sidechains due to the removal of contacts with neighboring proteins in the crystal, and the replacement of most of the intramolecular hydrogen bonds involving sidechains by hydrogen bonds to solvent molecules. For another small protein, APP, a fairly abrupt reorientation of a Phe sidechain ($\Delta\chi^1 \approx 90°$) in the β loop region is followed by a widening of that loop and replacement of intramolecular hydrogen bonds in the loop by bonds to solvent.

Trypsin is a much larger protein than BPTI or APP, so that the simulation of this enzyme in enough explicit water (4785 molecules) to approximate dilute solution conditions was greatly facilitated by access to a supercomputer (Wong & McCammon, 1986*a,b*). The simulation confirms previous results as far as general characteristics of protein motion are concerned, but also allows detailed studies of solvation and inhibitor binding. The average rms atomic position fluctuation for the protein atoms during 19 ps at 300K is 0.068 nm. The variation of fluctuation amplitude

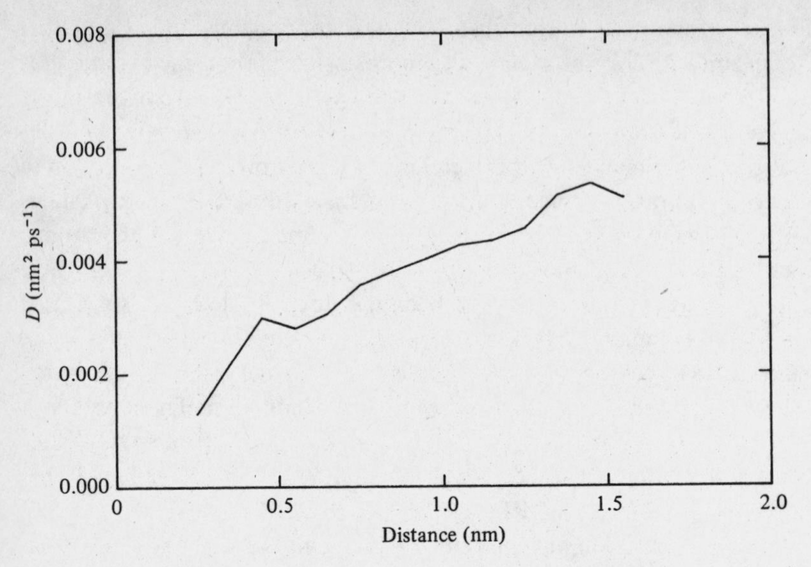

Fig. 5.2. Mean diffusion constants of the water molecules as a function of their starting distances to the closest atoms of the enzyme trypsin. The solvent at the protein surface is significantly less mobile than pure solvent.

with distance from the protein center is very similar to that already described for cytochrome c; the average amplitude is close to 0.05 nm in the interior, but increases rapidly to 0.1 nm for distances increasing from 1.5 to 2.7 nm (i.e., in the region of the more loosely packed surface). Other structural correlations as well as the anisotropy of the protein atom motions are also similar to what has been described above. The average mobility of the solvent depends strongly on distance from the protein, as shown in figure 5.2. The effective diffusion constant for water at the protein surface is less than half the bulk value ($D = 0.0036$ nm^2 ps^{-1}) for the SPC water model used in the simulation. This reduced mobility probably reflects simple geometric factors (the protein acts as a barrier to water displacement) as well as specific interactions (such as solvent hydrogen bonding to charged groups on the protein). Between 1.0 and 1.5 nm from the protein, the solvent diffusion constant is somewhat larger than the bulk value. This could correspond to a relatively disordered region of solvent that results from competition between the structural influence of the protein and its solvation shells on the one hand, and the forces that determine bulk water structure on the other. This simulation has also been useful in studies of solvation effects on the binding of enzyme inhibitors (cf. chapter 9).

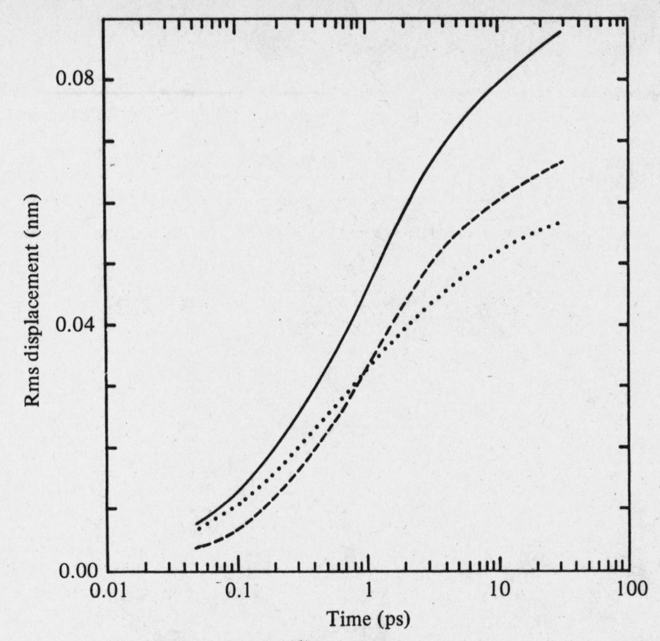

Fig. 5.3. Rms displacements in cytochrome c, averaged over different time intervals in a molecular dynamics simulation (Morgan *et al.*, 1983). Results are given for the net atomic displacements (——), the displacements of residue centroids (– – –), and the displacements of atoms relative to their residue centroids (. . . . . .). Collective motion, as gauged by the centroid displacements, dominates the net atomic motion for time intervals longer than 1 ps. (Copyright © 1983, John Wiley & Sons, Inc. Reprinted by permission.)

5.2.2 *Collective aspects*

Several of the results described in the previous subsection indicate that collective motions are of importance in protein dynamics. For example, collective motions are suggested by the collinearity of the preferred axes of displacement of a number of atoms in figure 5.1(*a*). That such collective components are present is consistent with the large amplitude of the atomic position fluctuations in the closely packed structure of the protein; large amplitude displacements must involve the collective motion of an atom and its neighbors. A number of studies have focused on these collective motions.

In one study, the atomic displacements in a dynamical simulation of cytochrome c were decomposed into two parts (Morgan *et al.*, 1983). The first part is the displacement of an atom relative to the centroid of the amino acid residue to which it belongs, and the second part is the displacement of the residue centroid itself. The former component largely

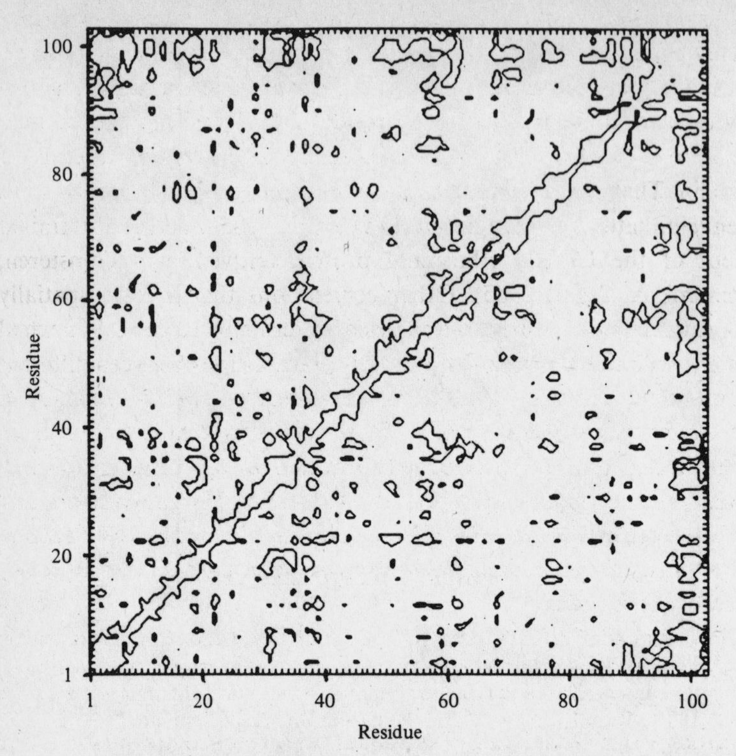

Fig. 5.4. Map showing the extensive residue displacement correlations in a 32 ps dynamical simulation of cytochrome c (McCammon, 1984). The correlation of residues i and j is computed as the normalized quantity $C_{ij} = \langle R_i R_j \rangle / \langle R_i^2 \rangle^{\frac{1}{2}} \langle R_j^2 \rangle^{\frac{1}{2}}$, where R_i is the instantaneous displacement of the centroid of residue i from the mean position of the centroid. The values of C_{ij} range from 0.6 to 1.0. A single contour is drawn at the value 0.87; larger correlations occur along the diagonal and within the small closed regions off the diagonal.

reflects localized motions, while the latter component is a probe of the collective motion. This study showed that the local components dominate the magnitude and directionality of the net atomic displacements during time intervals less than 1 ps. Over longer time intervals, the collective component tends to dominate the atomic displacements. These results are illustrated in figure 5.3. The backbone displacements and their variation throughout the tertiary structure are strongly dominated by collective dynamics over these longer time intervals. The sidechains tend to ride along with the backbone, but have additional local mobility.

The spatial extent of the collective motions in cytochrome c has been probed by calculating the displacement correlations of the residue centroids (McCammon, 1984). The results, displayed in figure 5.4, show that many residues have large cross-correlations. Some of these correlations are easily understood. For example, the correlated motion of residues 92–103 reflects collective motion of the C-terminal alpha helix, and the correlations of these residues with residues 1–10 reflects the contact between the C-terminal and N-terminal helices in the native structure of the protein. That alpha helices can move as semirigid bodies is also clear in recent simulations of myoglobin (Levy *et al.*, 1985) and the C-terminal fragment of the L7/L12 ribosomal protein (Aqvist, van Gunsteren, Leijonmarck & Tapia, 1985). Other correlations that involve spatially distant residues may be best rationalized in terms of large scale normal modes of oscillation or large scale structural transitions between different conformational substates (see below). In another study, based on a molecular dynamics simulation of BPTI, it was shown that distinct regions of the protein could be identified within which groups of atoms display very similar displacement time correlation functions (Swaminathan *et al.*, 1982). This result suggests that collective motion dominates the displacements within each such region. These regions include atoms from as many as seven or more residues.

The importance of collective motion in BPTI has also been demonstrated by normal mode calculations on the protein (Gō *et al.*, 1983). In this study, it was shown that most of the low frequency modes involve concerted displacement of many atoms. For modes with frequencies in the range $\nu = 3.6$–$6\ ps^{-1}$ (120–200 cm^{-1}), the motions typically involve several neighboring residues in the three dimensional structure. For modes with frequencies below 3.6 ps^{-1} (120 cm^{-1}), the motions are more global in character. The patterns of collective displacement in two of these low frequency modes are illustrated in figure 5.5. The low frequency motions make the dominant contribution to the net atomic displacements. In the calculations described by Gō *et al.* (1983), only the dihedral angles were allowed to vary. Brooks & Karplus (1983) carried out a normal mode analysis of BPTI in the full configuration space. The normal mode spectrum is shifted to lower frequencies, with about 60 modes below $\nu = 1\ ps^{-1}$ (33 cm^{-1}) versus the 36 found by Gō *et al.* This downward shift is as expected, because the freezing of bond angles (which is required for the simulation in dihedral angle space) is known to produce a molecular model that is somewhat too rigid (van Gunsteren & Karplus, 1982*a*). The important conclusions are, however; unchanged, the atomic displacements are generally dominated by modes with frequencies below 1 ps^{-1} and these modes are mostly delocalized over the protein. Levitt *et al.* (1985) have determined and extensively analyzed the normal modes of several proteins, again taking the dihedral angles as variables. Presumably because they

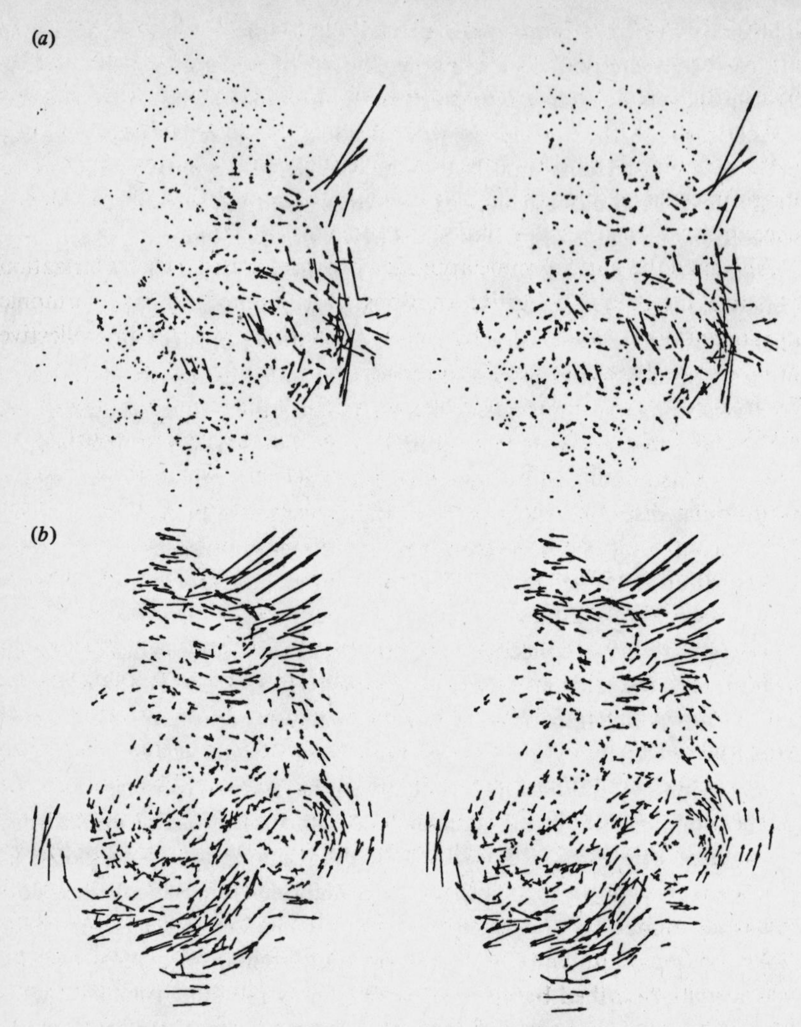

Fig. 5.5. Stereoviews of the directions of atomic displacement in two low frequency vibrational modes of the pancreatic trypsin inhibitor, as determined by a normal mode calculation (Gō *et al.*, 1983). Results are given for modes at frequencies (*a*) $\nu = 3.56\ \mathrm{ps}^{-1}$, (*b*) $\nu = 0.21\ \mathrm{ps}^{-1}$.

used a potential function that was intrinsically softer than those used in the studies described above, Levitt *et al.* (1985) obtained a normal mode spectrum for BPTI similar to that obtained by Brooks & Karplus (1983). The general conclusions are, however, in accord with those of the above studies. Several new findings from the work by Levitt *et al.* (1985) are of interest. For lysozyme and ribonuclease, it was found that motions that open and close the active site clefts are prominent in the low frequency modes. For all of the proteins, the mean square atomic displacements obtained by superposition of the normal modes are in reasonable agreement with those determined from the X-ray diffraction temperature factors (cf. section 5.2.4). Finally, similar normal modes were found for two structures of BPTI, namely the X-ray structure and one obtained after a long molecular dynamics simulation, despite the fact that the rms difference in atom positions for these structures was 0.25 nm.

Although the normal mode approach provides a useful characterization of some important collective motions, it is limited by the harmonic approximation. Molecular dynamics simulations show that collective motions are generally rather anharmonic in character (cf. section 5.2.1). Of greatest interest in this regard are transitions that involve the concerted shifting of groups of atoms as the protein crosses free energy barriers to move among slightly different stable conformations. Such transitions are responsible for the multimodal position distributions observed for individual atoms, as described above. They are also likely to make a significant contribution to the thermodynamic properties of proteins (Karplus & McCammon, 1981*a*). Such transitions have been observed to occur reversibly in dynamical simulations of cytochrome c (Mao *et al.*, 1982*a*) and BPTI (Levitt, 1983*b*). In the cytochrome c case, several transitions involving clusters of residues in the native structure were observed during a 32 ps simulation. The transitions occurred in less than a picosecond, and the protein occupied each conformational substate for several picoseconds. In the BPTI case, several transitions involving more widely separated residues were observed during a 132 ps simulation. One particularly interesting transition involved a rearrangement of hydrogen bonds at both ends of a strand of beta sheet; only small atomic shifts occurred during this transition, which was reversed after a period of 40 ps. Reversible, collective transitions among native-like conformations are probably a general feature of protein dynamics. Longer simulations can be expected to reveal more transitions of this kind, because larger free energy barriers can be crossed and larger regions of configuration space can be explored.

5.2.3 *Dynamic models*

As mentioned in section 3.1, a useful first step in the quantitative analysis of motions in proteins is the application of phenomenological models. An analysis of this kind has been carried out for the torsional librations of buried tyrosine residues observed in a molecular dynamics simulation of BPTI (McCammon *et al.*, 1979). The librations examined are those about the axis passing through the gamma and zeta carbon atoms

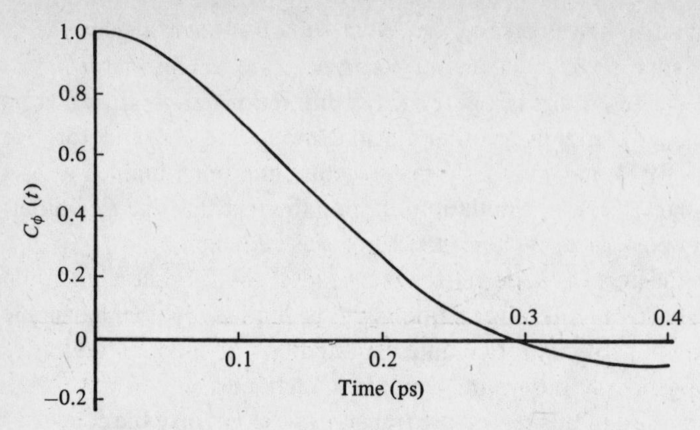

Fig. 5.6. The normalized time correlation function, $C_\phi(t) = \langle \Delta\phi(t)\,\Delta\phi(0)\rangle / \langle \Delta\phi(0)\,\Delta\phi(0)\rangle$, for torsional fluctuations $\Delta\phi$ of the tyrosine 21 ring in a molecular dynamics simulation of the pancreatic trypsin inhibitor (McCammon *et al.*, 1979). (Copyright © 1979, American Chemical Society. Reprinted with permission.)

of each ring (cf. figure 2.3). The motion is analogous to rotation of a benzene molecule about an axis passing through two atoms at opposite vertices of the ring. Within the protein, the amplitude of libration is limited by steric hindrance between atoms in the ring and those in the cage surrounding the ring. A typical amplitude observed at 300 K is 12°. The time correlation functions for the torsional fluctuations decay to small values in a short time (approximately 0.2 ps); subsequent decay occurs on a time scale of several picoseconds. This result, illustrated in figure 5.6, indicates that the torsional motion is predominantly local in character, with the ring rattling in its cage. On the longer time scale, collective distortions of the protein produce small changes in the overall shape of the cage with a concomitant alteration of the ring orientation. From the probability distribution of observed torsional librations $p(\phi)$, one can calculate the potential of mean torque $W(\phi)$ for the ring librations:

$$W(\phi) = -k_B T \ln p(\phi) \tag{5.2}$$

The potential $W(\phi)$ is roughly quadratic, which suggests that the ring libration can be analyzed in terms of the Langevin equation for a harmonic oscillator. This equation is

$$I\ddot{\phi} = -k\phi - f\dot{\phi} + N(t) \tag{5.3}$$

where $\phi(t)$ is the torsional displacement, I is the moment of inertia of the ring about the torsional axis, f is a friction constant, k is a harmonic restoring force constant obtained from the potential of mean torque, and

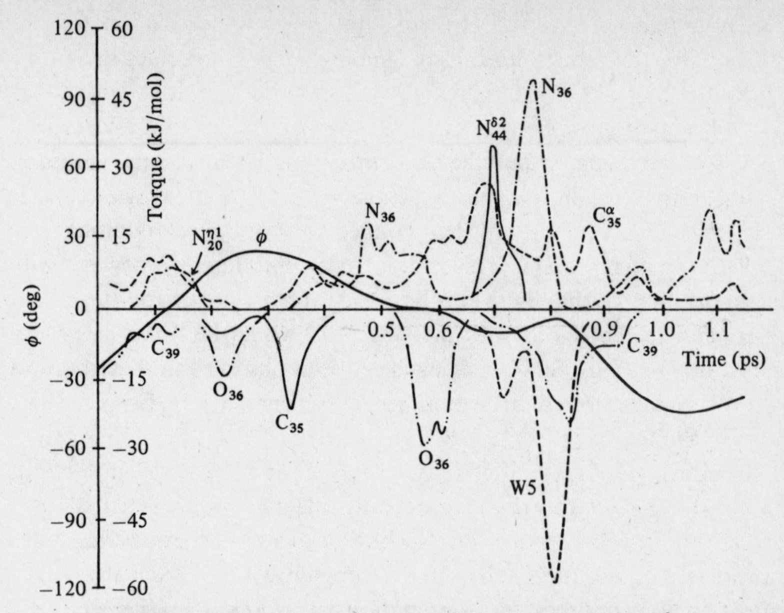

Fig. 5.7. Torques exerted on the tyrosine 35 ring due to nonbonded interactions with the surrounding atoms in a molecular dynamics simulation of the pancreatic trypsin inhibitor (McCammon & Karplus, 1980*b*). (Copyright © 1980, John Wiley & Sons, Inc. Reprinted by permission.)

$N(t)$ represents the random torques acting on the ring due to fluctuations in its environment. The Langevin equation is appropriate if $N(t)$ varies more rapidly than $\phi(t)$; in this limit, $N(t)$ may be modeled as a Gaussian random process and it is not necessary to specify the mechanism by which the torque fluctuations arise. This condition is approximately satisfied for the tyrosine rings; the duration of and interval between typical collisions involving the ring atoms and the surrounding cage atoms are of the order of 0.1 ps, which is somewhat shorter than the ring libration correlation time (cf. figure 5.7). The moment of inertia is readily calculated from the known atomic masses and geometry of the ring, and the force constant k is known from the shape of the potential of mean torque. The friction constant f can be calculated from the decay time τ of the torsional fluctuation correlation function by the relation $\tau = f/k$. The numerical values of these three coefficients indicate that the rotational motion of the ring within its cage is nearly critically damped; this is consistent with the shape of the time correlation function in figure 5.6. The friction constant may be related to an angular diffusion constant by use of the Einstein formula $D = k_B T/f$. The calculated value, $D = 2 \times 10^{11}\ s^{-1}$, is of the same order of magnitude as the experimental diffusion constants for the corresponding rotational motions of small aromatic molecules in organic solvents (e.g., $D = 7.9 \times 10^{10}\ s^{-1}$ for benzene in isopentane). Physically, this result makes sense because the protein rings examined are located in hydrophobic regions within the protein. That the rotational diffusion constant in the protein is somewhat larger than that in a typical organic solvent may be due to the covalent connectivity of the surrounding cage atoms within the protein. This connectivity interferes somewhat with efficient packing, so that there may be fewer matrix atoms in positions that lead to effective collisions with the ring than would be found in typical solvent surroundings. Similar effects have been observed in spectroscopic studies of small molecules in large molecule solvents (Moog, Ediger, Boxer & Fayer, 1982).

The above analysis is restricted to motion on a time scale of a few tenths of a ps. It is only for these short intervals that the calculated time correlation functions are accurate according to strict criteria (Zwanzig & Ailawadi, 1969). An attempt has been made, however, to extend the above analysis to longer times by examination of atomic displacement time correlation functions over periods of up to 10 ps (Swaminathan *et al.*, 1982). A number of interesting qualitative conclusions emerged from this work. Dynamical simulations of BPTI in the presence and absence of a simple nonpolar solvent environment were found to yield similar rms atomic displacements. Time correlation functions for the atomic displacements decay somewhat more slowly in the presence of solvent, particularly for atoms near the surface of the protein. In the vacuum calculation, the range of correlation times was 0.2–5 ps whereas the solvent calculation results ranged from 0.4 to 10 ps. For the protein in solvent, the correlation times for atoms in the exterior tend to be substantially larger than those for atoms in the interior of the protein. Effective friction constants for the atoms were estimated using a simple harmonic oscillator Langevin equation approach. From these friction constants, it was estimated that the collective motions that dominate the atomic displacements over long times typically involve structural units containing of the order of 100 heavy atoms.

Applications of the Langevin equation allow one to identify some of the important contributions to the dynamics of particular motions in protein molecules. In principle, a more detailed decomposition of the dynamics is possible by microscopic models of the type employed in kinetic theory. In these, one attempts to replace the frictional and random force of Langevin-type models by specific mechanical models that recognize, e.g., the masses and collision geometries of the atoms in the system. Such a microscopic analysis has been attempted for the subpicosecond ring librations that were described above (McCammon *et al.*, 1979). Because the ring environment bears some resemblance to an organic solvent, the microscopic model employed is of the Enskog type that has been used successfully in the study of simple liquids (Hynes, 1977). In this model, the reorientation of a molecule in a liquid is assumed to occur as a result of successive binary collisons between the repulsive van der Waals cores of neighboring molecules. The successive collisions are assumed to be uncorrelated and of instantaneous duration and to randomize the angular velocity of the reorienting molecule; this molecule moves freely between collisions. For the tyrosine ring in the protein, the model was modified to take into account the fact that, between collisions, the ring moves in a harmonic manner due to the torsional restoring force. In the absence of collisions, the ring reorientation is described by a simple harmonic oscillator equation of motion

$$I\ddot{\phi} = -k\phi \tag{5.4}$$

where it is assumed that $\langle\phi\rangle = 0$. The unperturbed oscillator exhibits undamped librations at a frequency $\omega_0 = (k/I)^{1/2}$. The normalized time correlation function for the torsional angle is simply

$$C_\phi^0(t) = \cos \omega_0 t \tag{5.5}$$

If the oscillator is subject to instantaneous, uncorrelated collisions, and if each collision randomizes the angular momentum of the oscillator, then the displacement time correlation function of the perturbed oscillator is (McCammon *et al.*, 1979)

$$C_\phi(t) = \exp(-\nu t/2)[\cos at + (\nu/2a)\sin at] \tag{5.6}$$

where ν is the reciprocal of the mean time between collisions and $a^2 = \omega_0^2 - (\nu^2/4)$. With reasonable parameters, this model successfully reproduces the simulation time correlation function for periods of several tenths of a picosecond. The simple microscopic model is, however, inadequate in several respects. As can be seen from figure 5.7, most of the collisional torques are not sufficiently strong to randomize the angular velocity of the ring. Also, several impulses in rapid succession are sometimes observed for particular ring–atom, matrix–atom pairs (most commonly when the matrix atom is part of the local backbone). It would be desirable to incorporate these features into a more satisfactory microscopic model.

5.2.4 *Experimental connections*

A variety of comparisons between simulation and experimental studies of protein dynamics have been made. In general, the pictures that result from the two types of study are in agreement. Because the simulations and the analyses of experimental studies both involve approximations, differences in certain detailed results are inevitably found. Approximations in the simulations include those in the underlying potential functions (e.g., neglect or simplified description of the solvent surroundings, as well as approximations in the potential functions used for the proteins themselves), those arising from the use of classical rather than quantum dynamics, and those associated with the finite lengths of the simulations. Approximations in the experimental work largely reside in the models that are used to interpret the raw data in terms of molecular motion, although uncertainties also exist in the data due to finite sampling, tentative assignments of spectral lines, etc. Thus, both the theoretical and experimental work stand to benefit from critical comparisons of their results. Where the results agree, one obtains greater confidence not only in the combined picture of molecular motion but also in pictures of analogous motions that may have been obtained by only one type of approach. On the other hand, disagreements can lead to improved methods for simulation and/or experimental analysis, as will be described in chapter 9. An excellent account of some of these topics has been presented recently by Levy & Keepers (1986).

X-ray diffraction studies provide information on the spatial distribution of electron density in protein crystals. The method has been extensively applied to determine the average positions of the atoms in protein molecules from the corresponding peaks in the electron density. As described in chapter 4, these positions are generally used to begin dynamical simulations of proteins. The results of conventional X-ray studies are averaged both over the time required to collect the diffraction data (usually at least several hours) and over the molecules in one or more crystals from which the data are collected. Although this averaging ordinarily eliminates the possibility of obtaining time-dependent information, a variety of important connections can be drawn between the experimental and theoretical results (Northrup *et al.*, 1980; Petsko & Ringe, 1984; Levy & Keepers, 1986; Ringe & Petsko, 1985). The mean positions obtained in a simulation can be compared with those from the X-ray studies, as can several average measures of the thermal displacements of the atoms. As noted previously, most molecular dynamics simulations of proteins to date have been carried out without attempting to include the crystalline environment of the protein. The protein potential function has usually been modified to account roughly for dielectric effects in the largely aqueous surroundings, but this is not sufficient to represent the detailed interactions with neighboring water and protein molecules in a crystal. Despite these limitations, the time average structures obtained in most simulations are similar to the X-ray structures. Typical rms differences between average atomic positions in simulations and X-ray structures are in the range 0.2–0.3 nm; differences in the interior tend to be smaller than those on the surface of the protein. In part, this agreement results from the short length of the simulations; there is not sufficient time for the molecule to denature in response to the absence of hydrophobic and other effects. This is clear from the tendency of most proteins to move progressively if slowly away from the X-ray structure during long simulations (Levitt, 1983*a*, *b*; Levy *et al.*, 1985). While the general similarity between the computed and experimental structures may thus represent a fortuitous metastability in most simulations, it does mean that the atoms in the protein interior move in realistic environments.

In the simplest case of isotropic, harmonic atomic displacements, the widths of electron density peaks are characterized experimentally by a single Debye–Waller factor (or temperature factor), B, for each atom. This factor is related to the apparent mean square atomic displacement by (Willis & Pryor, 1975)

$$B = (8\pi^2/3)\langle(\Delta\mathbf{r})^2\rangle \tag{5.7}$$

The apparent mean square atomic displacement can be decomposed into a term $\langle(\Delta\mathbf{r})^2\rangle_{\mathrm{ld}}$ from lattice disorder in the crystal and other effects, plus a term $\langle(\Delta\mathbf{r})^2\rangle_{\mathrm{cv}}$ for thermal motion (conformational and vibrational fluctuations) within a single molecule. To distinguish between these contributions, Debye–Waller factors can be measured at two or more different temperatures, since lattice disorder should be relatively insensitive to temperature, while $\langle(\Delta\mathbf{r})^2\rangle_{\mathrm{cv}}$ should depend on temperature. In practice, a variety of complications make it difficult to rigorously separate these two terms. As a protein crystal is cooled, transitions between internal structural substates will be slowed so that, relative to a given time scale, some conformational fluctuations may freeze into static disorder. In addition, the reduction of mean amplitudes of atomic motion with decreasing temperature may reflect both the reduced amplitude of low frequency vibrations and the depopulation of certain substates by transitions to substates of lower energy. The distribution between frozen-in disorder and annealing clearly depends on the rate of cooling. Finally, there is the possibility that the disorder term may reflect temperature sensitive vibrations involving the concerted motions of more than one protein molecule

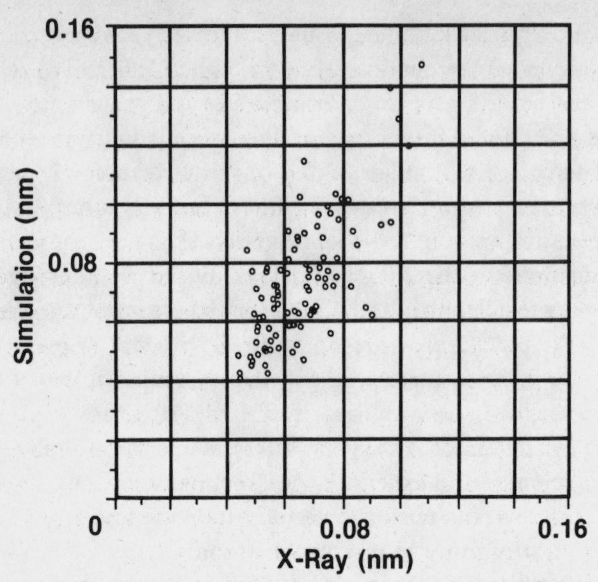

Fig. 5.8. Correlation diagram for the calculated and experimental values of the average rms atomic position fluctuation for each residue in cytochrome c (Northrup *et al.*, 1981). The correlation coefficient is $R = 0.73$.

(Finzel & Salemme, 1985). Complications also arise in the detailed interpretation of the two terms at any given temperature. The conformational–vibrational term is known to vary somewhat with the crystal structure; contacts between groups from neighboring proteins generally lead to reduced amplitudes of motion for the groups involved. Sheriff *et al.* (1985) have introduced a method to correct approximately for the effects of such contacts, so that a better description of the intrinsic flexibility of a protein can be obtained from the crystal data.

Despite the potential difficulties indicated above, useful comparisons have been made between the atomic displacements inferred from X-ray data and those obtained from molecular dynamics simulations. Such comparisons depend on the disorder contribution being fairly constant throughout the protein, so that it can be subtracted out (Frauenfelder, Petsko & Tsernoglou, 1979; Northrup *et al.*, 1980) or treated as a constant baseline. In general, the possibility of nonuniform disorder must be considered (Artymiuk *et al.*, 1979; Finzel & Salemme, 1985). The assumption of uniform disorder appears to be a reasonable one in some cases, however; e.g., for myoglobin crystals, there is some evidence that static disorder due to rotational misalignment of the molecules is small (Fiamingo, Thorkildsen & Brill, 1980). In such cases, the small apparent variation of the disorder term allows one to estimate $\langle(\Delta\mathbf{r})^2\rangle_{\mathrm{cv}}$ and make comparison with dynamical simulation results. Comparisons based on a 32 ps simulation of cytochrome c (Northrup *et al.*, 1980, 1981) and on a series of 25 ps simulations of myoglobin (Levy *et al.*, 1985; Levy & Keepers, 1986) showed reasonably good agreement between the theoretical and experimental results; cf. figure 5.8. The simulations may even provide a more accurate description of the atomic fluctuations than is implied by the above comparisons. It has recently been shown that the standard techniques for interpreting X-ray data tend to underestimate the amplitudes of atomic displacement, especially for those atoms with the largest displacements (Levy & Keepers, 1985; Levy *et al.*, 1986). These techniques produce distortions similar to the pattern seen in figure 5.8, in part because they incorporate the assumption of harmonic atomic motion. The agreement between simulation and X-ray results is perhaps surprising in view of the different time scales involved. While proteins clearly do not sample all of the similar conformational states that make up the native structure during periods of a few dozen ps (cf. section 5.2.2), it may be that they can sample a fairly representative subset of these states. More importantly, the normal mode studies described in section 5.2.2 make clear that low frequency structural vibrations (as distinct from structural transitions) account for a sizable part of the atomic fluctuation magnitudes. Since the lowest frequencies encountered (often about 3 cm^{-1}) correspond to motions with periods of about 10 ps, any simulation longer than this can be expected to sample most of the vibrationally accessible space, even if the sampling of the space accessible by transitions is incomplete.

Some effort has been made to go beyond the simple assumptions of isotropic and harmonic motion in the analysis of experimental results. These include X-ray structure refinements with partial or full analysis of the atomic displacement anisotropy (Artymiuk *et al.*, 1979; Konnert & Hendrickson, 1980; Glover *et al.*, 1983). The results obtained to date are consistent with the theoretical finding that the atomic motion is, in general, quite anisotropic. Experiment and theory are also in agreement in concluding that collective motions within proteins tend to wash out the correlations that might be expected between local structure and the preferred direction of atomic displacement. Studies of the temperature dependence of the electron density peak widths provide valuable information on the character of the effective potentials constraining the atoms to their mean positions (Frauenfelder *et al.*, 1979; Hartmann *et al.*, 1982). Such studies show that the atomic displacements are generally anharmonic in character. Frauenfelder *et al.* (1979) have characterized the effective atomic potentials by simple power law expressions. Among other findings, they show that the potentials of mean force for atomic displacement near the protein surface often have a square well character, in agreement with theoretical findings. Van Gunsteren *et al.* (1983) carried out a 20 ps simulation of a full unit cell containing four BPTI molecules along with solvent water molecules, and showed that the expected diffraction from this model crystal was in reasonable accord with experimental results. This pioneering study makes clear that the interpretation of X-ray data can be improved by using information from simulations on the anisotropic and anharmonic nature of atomic motions. This point is discussed further in chapter 9.

Nuclear magnetic resonance is particularly valuable as a probe of protein dynamics. The method can be used on the normal solution state of proteins and, as a result of recent developments in spectral assignment techniques, can provide structural and dynamical information on specific parts of protein molecules. Moreover, the method is useful for study of both fast and slow local motions. Fast motions modulate the magnetic environment of nuclei at frequencies that stimulate nuclear magnetic relaxation. Thus, these fast motions are reflected in NMR relaxation times and in the nuclear Overhauser enhancement (NOE) factors that characterize the cross relaxation of specific pairs of nuclei. Studies of ^{13}C nuclei have been particularly useful, because the magnetic field fluctuations are often dominated by the relative motion of protons that are bonded to the carbon with a well characterized geometry. The reorientation of C—H bonds occurs on several time scales (Levy, Karplus & McCammon, 1981). Partial reorientation occurs on the time scale of a few ps as a result of structural fluctuations of the kind described earlier in this chapter. Some additional reorientation occurs on a longer time scale as the protein samples different but similar conformational substates. For proteins in solution, complete reorientation occurs over times of a few nanoseconds as a result of overall rotational diffusion of the molecule. The extent of reorientation produced by internal motion can be characterized by an order parameter S^2 for each C—H bond,

$$S^2 = \frac{4\pi}{5\langle r^{-6}\rangle}\sum_m \left\langle \frac{Y_m^2}{r^3} \right\rangle^2 \tag{5.8}$$

Here, the angular brackets indicate a time average, r is the C—H bond length, and the Y_m^2 are the second-order spherical harmonic functions of the C—H bond orientation in a molecule-fixed frame. The quantity S^2 takes values ranging from zero to one; these limits correspond to isotropic motion and to no reorientation, respectively. In calculating order parameters from molecular dynamics simulations where explicit methyl hydrogen atoms have not been included, the order parameter of equation (5.8)

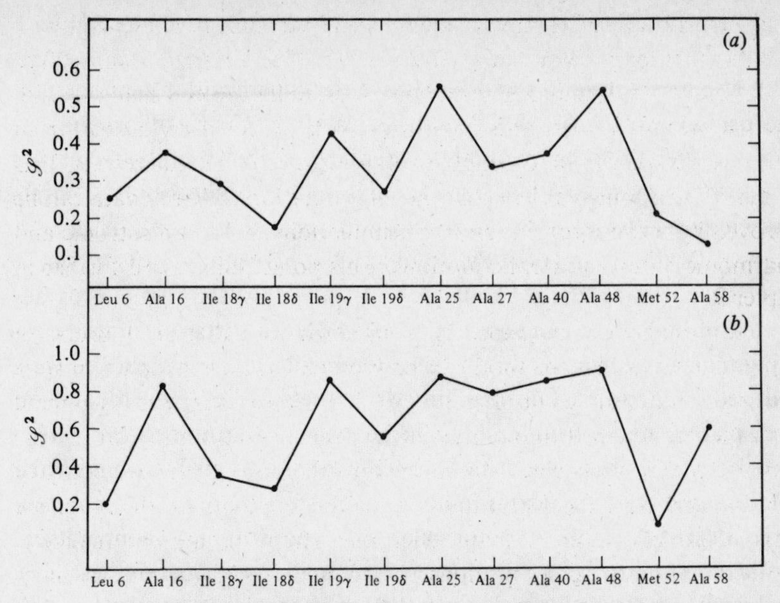

Fig. 5.9. Calculated and experimental NMR order parameters for the methyl groups in BPTI (after Lipari *et al.*, 1982). The order parameter S^2 is that for the axis defined by the methyl carbon and the heavy atoms to which it is joined. The experimental values are shown in (*a*). The values calculated from a 96 ps molecular dynamics simulation are shown in (*b*). Copyright © 1982, Macmillan Journals Ltd. Reprinted by permission.

can be decomposed into the product of (1) the contribution from rotations about the bond connecting the methyl group with rest of the molecule and (2) that due to motions of this bond itself (Levy *et al.*, 1985). The former is treated as a constant, while the latter can be calculated from the simulation.

Lipari, Szabo & Levy (1982) have computed order parameters for twelve methyl groups from a 96 ps simulation of BPTI and have compared these to values of S^2 extracted from NMR relaxation data (Richarz, Nagayama & Wüthrich, 1980; Lipari & Szabo, 1982). The results are displayed in figure 5.9. While experiment and simulation are largely in agreement with regard to the relative mobility of the methyl groups, the differences in magnitude indicate that larger amplitude motions occur in the experimental system. Analyses of a more recent 132 ps simulation of BPTI (Levitt, 1983*b*) and of a 300 ps simulation of myoglobin (Levy *et al.*, 1985) also indicate that methyl groups of real proteins in solution undergo larger reorientations during a few ns than are observed in simulations on a 100 ps time scale; that is, the molecular dynamics order parameters are not fully converged in 100–300 ps.

While X-ray diffraction and NMR are the experimental methods that have provided the most detailed structural descriptions of protein dynamics, a variety of other methods have either provided useful information or hold some promise of doing so. There have been a number of attempts in recent years to use Mössbauer spectroscopy as a probe of protein dynamics (Knapp, Fischer & Parak, 1982; Bauminger *et al.*, 1983; Nadler & Schulten, 1984; Parak & Knapp, 1984). This technique provides information on the mean square displacements of certain atomic nuclei during intervals corresponding to the lifetime of the nuclear excited state produced by gamma ray absorption. Most investigations have focused on the displacements of iron atoms in heme proteins during periods of 10^{-7} to 10^{-9} s. Partly because there is usually only one probe atom in the protein, some difficulty arises in any attempt to specify a unique microscopic description of the molecular motion, although temperature-dependent studies help to reduce the number of possibilities. The conclusions emerging from these studies emphasize the dominant contribution of collective motions within the proteins to the iron displacements. The motions appear to include transitions among different substates. For a number of proteins, the apparent displacements decrease dramatically below certain critical temperatures (200 K for metmyoglobin and deoxymyoglobin); this has been attributed to a freezing of the water at the protein surface and concomitant suppression of the important collective motions. Because the Mössbauer technique is not sensitive to low frequency motions, it has been useful in the approximate separation of the static and dynamic contributions to the atomic displacements observed in X-ray diffraction studies (Frauenfelder *et al.*, 1979; Hartmann *et al.*, 1982).

Many of the hydrogens bonded to oxygen or nitrogen atoms in a protein will be replaced by deuterium or tritium if the molecule is suspended in water that contains these isotopes of hydrogen. Such exchange requires transient breaks in any hydrogen bonds at the exchange site and the presence of solvent species at this site. The hydrogen exchange phenomenon has long been recognized as an indicator of conformational fluctuations in proteins (Englander, Downer & Teitelbaum, 1972; Woodward & Hilton, 1979; Englander & Kallenbach, 1984). The detailed character of the conformational fluctuations has been the subject of some debate, in part due to the difficulty of identifying which hydrogens are exchanged in a given period of time. Applications of proton NMR methods to this problem, together with advances in NMR assignment techniques, offer some hope of making hydrogen exchange a more detailed tool for probing protein dynamics (Hilton & Woodward, 1978; Wüthrich & Wagner, 1979; Wagner & Wüthrich, 1979, 1982; Griffey, Redfield, Loomis & Dahlquist, 1985). The NMR techniques have more recently been complemented by neutron diffraction techniques which indicate the locations of exchanged hydrogens in protein crystals soaked in heavy water (Kossiakoff, 1982, 1985; Wlodawer & Sjölin, 1982), and by chemical techniques for analyzing fragments of proteins following exchange (Englander & Kallenbach, 1984). Overall, the results of the recent studies are consistent in showing that the slowest exchange is associated with backbone hydrogens that are buried in the protein interior and centrally located in substantial elements of secondary structure. This is consistent with the observation by Levitt (1981) that there is some correspondence between the fluctuations in hydrogen bond lengths in dynamical simulations and the hydrogen exchange rates. Kossiakoff (1982) suggests that the slow exchange of buried hydrogens proceeds by a regional melting mechanism that involves the local breaking of hydrogen bonds and the formation of a solvent-filled cleft to the protein surface. This model is intermediate between the traditional penetration and local unfolding models, which suggest, respectively, less and more disruption of the protein matrix. A detailed discussion of the various models and the kinetic signatures by which they might be distinguished has been given by Englander & Kallenbach (1984). These authors also indicate several cases in which the local unfolding mechanism appears to be operative. In one case, the peptide hydrogens involved in the hydrogen bonds of the S peptide alpha helix of ribonuclease S all exchange at similar rates, despite the fact that some of these bonds are at the protein surface and some are buried in the interface between the S peptide and the rest of the protein. The buried hydrogens would be expected to exchange more slowly than the exposed ones in any simple solvent penetration model, while the uniform rates observed could correspond to a cooperative unwinding of the helix. In another case, classes of peptide hydrogens that are kinetically similar, and whose exchange rates respond similarly to allosteric transitions, have been shown to be associated with individual helix segments in hemoglobin. The local unfolding units identified so far appear to correspond to segments of secondary structure containing up to about ten residues. The unfolded lifetimes are estimated at somewhat less than 10^{-7} s; this is too long for direct comparisons with molecular dynamics, but could perhaps be compared with refined Brownian dynamics simulations (chapter 7). Although the dynamic interpretation of the hydrogen exchange results is still somewhat imprecise, an important finding in the recent work is that surprisingly rapid exchange is observed from some well buried atoms with small temperature factors in the X-ray diffraction data. Thus, despite the results of Levitt (1981) mentioned above, it is not always possible to extrapolate from the relatively common, low energy motions measured by temperature factors (or rms displacements in conventional dynamical simulations) to the infrequent, high energy motions probed by hydrogen exchange.

It was shown a number of years ago that oxygen molecules can diffuse into the interior of proteins freely enough to quench the fluorescence of buried tryptophan sidechains on a nanosecond time scale (Lakowicz & Weber, 1973). The apparent diffusion constant for oxygen in the interior of a protein is nearly as large as that for diffusion in bulk water. Many subsequent studies (summarized by Calhoun, Vanderkooi, Woodrow & Englander, 1983*b*) have examined the quenching of tryptophan fluorescence and phosphorescence by other molecules and ions. Calhoun, Vanderkooi & Englander (1983*a*) and Calhoun *et al.* (1983*b*) have provided results which indicate that, while oxygen can move through native proteins fairly freely, ions and polar molecules larger than oxygen are strongly excluded from the interior of native proteins and depend upon substantial disturbances of the protein structure for tryptophan contact and quenching.

When the fluorescence from aromatic residues in a protein results from excitation by a pulse of polarized light, the subsequent reorientation of these residues can be probed by monitoring the resulting fluorescence depolarization. Recent developments in both experimental methods and data analysis have begun to allow comparison with the results of dynamic simulations (Kasprzak & Weber, 1982; Levy & Szabo, 1982; Ichiye & Karplus, 1983; Lakowicz & Maliwal, 1983; Szabo, 1984; Beechem & Brand, 1985; Levy & Keepers, 1986). Of particular interest here are the results reported by Kasprzak & Weber (1982) and Lakowicz & Maliwal (1983) concerning tyrosine ring reorientations in BPTI; these experimental results are consistent with the amplitudes and time scales of ring motion observed in dynamic simulations (section 5.2.1).

From the above discussion, it is clear that the X-ray diffraction and NMR studies provide the most information concerning the structural aspects of motions in proteins. Even in these studies, however, theoretical modeling has been helpful in the development of detailed structural interpretations. Theory is likely to play a larger role in the future with respect to the interpretation of these and other types of experiments (cf. chapter 9). Among the various experimental approaches, inelastic neutron scattering has perhaps the greatest unrealized potential. In the one comparison that is available, there is very little agreement between the scattering spectrum from BPTI in solution and the spectrum predicted by normal mode analysis (Smith *et al.*, 1985). In particular, none of the peaks predicted below 300 cm^{-1} appear in the comparatively broad and featureless experimental spectrum. This is likely due to a combination of anharmonic and solvent damping effects that are neglected in the normal mode analysis. Comparison of such experimental results with the results of molecular dynamics simulations of proteins in water will be of great interest. By appropriate isotopic labeling studies, the inelastic neutron scattering approach should be able to provide the same type of detailed dynamical information for proteins that it has for other dense materials (Leadbetter & Lechner, 1979; Ewen, 1984).

5.3 Results for nucleic acids

There was a delay of seven years from the first reported molecular dynamics simulation for a protein (McCammon, 1976) to the first reports on nucleic acid simulations (Levitt, 1983*d*; Prabhakaran, Harvey, Mao & McCammon, 1983; Tidor *et al.*, 1983). The results from the protein simulations had been cause for both hope and apprehension about the ability to use molecular dynamics to simulate motions in nucleic acids. Hope, because the simulations gave average protein structures close to the crystallographic structures and motions that were in reasonable agreement with thermal factors from crystallography (Northrup *et al.*, 1980; Levitt, 1983*b*). Apprehension, because the original protein simulations were done for models *in vacuo*. The inclusion of explicit solvent has been shown to improve the agreement with experiment (van Gunsteren *et al.*, 1983; Berendsen, van Gunsteren, Zwinderman & Geurtsen, 1986), and solvent effects should be even more important for nucleic acids, since the latter are polyelectrolytes (Eisenberg, 1976), and have larger surface-to-volume ratios than globular proteins.

The apprehensions turned out to be at least partly justified, because the nucleic acid models are sometimes prone to structural degradation when simulations are carried out *in vacuo* (Levitt, 1983*d*; Prabhakaran & Harvey, 1985). Structural stability can be improved by setting all electrostatic charges to zero (Levitt, 1983*d*); by scaling electrostatic charges and/or scaling the dielectric constant to mimic the effects of solvent and of counterion condensation (Prabhakaran, Harvey, Mao & McCammon, 1983; Tidor *et al.*, 1983; Singh *et al.*, 1985); by including explicit hydrogen bonds (Harvey *et al.*, 1985*a*); by the use of explicit solvent (Seibel *et al.*, 1986); and by the use of periodic longitudinal boundary conditions that make a DNA oligomer effectively a segment of an infinite double helix (Prabhakaran & Harvey, 1985).

The nucleic acid simulation that has been most extensively examined is

Table 5.1. *Molecular dynamics simulations of DNA*

Reference	Sequence	Comments
Levitt (1983*d*)	{[d(CGCGAATTCGCG)]$_2$ / dA$_{24}$·dT$_{24}$}	Electrostatic forces omitted; all hydrogens included
Tidor *et al.* (1983)	[d(CGCGCG)]$_2$	Normal mode analysis also reported; B and Z configurations
Singh *et al.* (1985)	d(CGCGA)·d(TCGCG)	With and without hydrated counterions
Seibel *et al.* (1985)	d(CGCGA)·d(TCGCG)	With solvent and counterions
Prabhakaran & Harvey (1985)	[d(CG)$_6$]$_2$	With and without intercalation site
Westhof *et al.* (1985)	d(Br^{5}C·G·Br^{5}C·G·Br^{5}C·G)	Z-DNA
van Gunsteren *et al.* (1986)	d(CGCAACGC)·d(GCGTTGCG)	With solvent and counterions

that of tRNA^{Phe} (Prabhakaran *et al.*, 1983; Harvey, Prabhakaran, Mao & McCammon, 1984; Prabhakaran, Harvey & McCammon, 1985*a*; Prabhakaran, McCammon & Harvey, 1985*b*; Harvey *et al.*, 1985*a*; Harvey, 1986). A number of double helical DNA oligomers have been simulated, and a list of all DNA molecular dynamics simulations of which we were aware as of December, 1985, is given in table 5.1. All of the tRNA and DNA studies have been done without explicit solvent except for those of Seibel *et al.* (1986) and van Gunsteren, Berendsen, Geurtsen & Zwinderman (1986), where full treatments of counterions and solvent were undertaken in dynamic simulations of DNA.

In this section, we present a summary of the information on the rapid motions (time scales less than about 10 ps) that has been derived from these molecular dynamics simulations on nucleic acids and from the normal mode analysis of DNA (Tidor *et al.*, 1983).

5.3.1 *Motions of individual atoms*

There are three principal quantities which we examine here, with the purpose of understanding the ways in which atomic motions of nucleic acids differ from those of globular proteins. First, we describe the amplitudes, comparing them to the thermal amplitudes observed in crystallographic studies and examining the relationship between thermal amplitude and solvent-exposed surface area. Second, we examine the anisotropies of atomic motions. Third, many of the motions produce distributions which differ markedly from those expected for harmonic motions, so we investigate these anharmonicities. The logical progression is from the lower moments to the higher moments of the probability distribution functions for atomic positions. The mean position of an atom is just the first moment, $\langle \mathbf{r} \rangle$; the mean square amplitude is derived from the second moment, $\langle (\Delta \mathbf{r})^2 \rangle$; the anisotropy is obtained by decomposing the second moment into its three principal orthogonal components (cf. section 5.2.1); and the anharmonicity is measured by the skewness and the kurtosis, which are derived from $\langle (\Delta \mathbf{r})^3 \rangle$ and $\langle (\Delta \mathbf{r})^4 \rangle$, respectively (cf. equation 5.1).

The amplitudes of the atomic motions observed in molecular dynamics simulations are dependent on the time interval over which the motions are measured (Morgan *et al.*, 1983; Prabhakaran *et al.*, 1985*a*). Over a time scale of tens of picoseconds, the reported root mean square atomic displacements for B-DNA have been in the range 0.03–0.08 nm (Tidor *et al.*, 1983) or 0.07–0.12 nm (Singh *et al.*, 1985), while those for tRNA are about 0.07–0.15 nm for all atoms except those in the single stranded 3′ CCA terminus (Prabhakaran *et al.*, 1985*a*). The reason for the difference in reported amplitudes in DNA is not clear, but the amplitudes of Tidor *et al.* (1983) are comparable to those observed in several protein simulations (Northrup *et al.*, 1981; Levitt, 1983*b*). The average amplitude in the tRNA simulation, 0.143 nm, compares very favorably with the value of 0.129 nm derived from the average crystallographic temperature factor ($B = 43.5$ Å^2; B. Hingerty, personal communication). The rms amplitudes in Z-DNA are considerably smaller (Tidor *et al.*, 1983; Westhof *et al.*, 1986), reflecting the greater rigidity of the Z conformation (Wang *et al.*, 1979; Ramstein & Leng, 1980). All of the foregoing simulations were done *in vacuo*. The addition of explicit solvent to a DNA simulation reduced the average amplitude to about one-half the value obtained *in vacuo* (Seibel *et al.*, 1985).

As important as the amplitudes themselves are the relative mobilities of various groups of atoms. In all of the nucleic acid simulations where relative mobilities have been reported, it has been observed that the phosphate atoms have the highest mobilities, the atoms of the ribose (or deoxyribose) have intermediate mobilities, and the base atoms have the smallest mobilities (Prabhakaran *et al.*, 1983; Tidor *et al.*, 1983; Singh *et al.*, 1985; Westhof *et al.*, 1986). This is in agreement with the relative mobilities derived from crystallographic thermal parameters of $tRNA^{Phe}$ (Sussman *et al.*, 1978) and with normal mode calculations on DNA (Tidor *et al.*, 1983). The same relative thermal parameters are also observed in

B-DNA crystal structures, but the initial interpretation in terms of dynamic disorder arising from thermal motion (Drew *et al.*, 1981) was later revised. The persistence of large B factors down to 16 K led to the conclusion that static disorder, not motion, was responsible (Drew, Samson & Dickerson, 1982). Of course, certain types of static disorder at low temperature may become dynamic at higher temperature. In the case of the tRNA simulation, the relative mobilities of each of the residues correlate very well with the average crystallographic thermal parameters for each residue (Prabhakaran *et al.*, 1985*a*); the correlation coefficient is 0.64, which is highly statistically significant ($P < 0.001$, $N = 76$). Apparently, the differences in atomic mobility are a consequence of simple packing considerations, because the vibrational amplitudes are highly correlated with solvent-exposed surface area (Harvey *et al.*, 1984).

The anisotropy of atomic motion is measured by the shape of the atomic thermal ellipsoid. (Isotropic motion produces a spherically symmetric probability distribution function.) For atoms in nucleic acids, both the shape and the orientation of the thermal ellipsoids are generally dependent on the time interval over which the data are analyzed. Although this time dependence is also observed for proteins (Morgan *et al.*, 1983), the underlying structural causes are different because of the fundamental differences in the structures of proteins and nucleic acids. In the double helical regions of tRNA, for instance, atoms in stacked bases have rapid, small amplitude motions along the helix axis, while the transverse motions are of larger amplitude but occur over a longer time scale. This is seen most clearly in moving pictures taken from computer graphics representations of successive structures from the simulation (Harvey, Prabhakaran, Suddath & McCammon, 1985*b*). When the movie is filmed so that 1 s of real time corresponds to about 1 ps of the simulation, one sees that the bases in double helical regions show very rapid motions along the helix axis. The relatively free rotation about the glycosidic bond allows each base to flutter and jitter in the cavity formed by its neighbors. Superimposed on this motion are the much slower transverse motions that require the coordinated translation of the base, the ribose, and part of the backbone. On time scales shorter than about 0.5 ps it is the longitudinal motions that predominate, while the dominance of the transverse motions only becomes evident over intervals longer than about 2 ps. Another similarity in anisotropies of atomic motions in tRNA and proteins has to do with the orientations of the thermal ellipsoids for those atoms that are covalently bonded to only one other atom. The preferred directions of displacement for these atoms are those that minimize the stretching of the single covalent bond. This effect is particularly pronounced over short time intervals; it does persist for more than 20 ps, although slow collective motions gradually dilute the effect (Morgan *et al.*, 1983; Prabhakaran *et al.*, 1985*a*).

As the previous paragraph shows, there are direct correlations between the anisotropies and local structural features. The search for patterns that might relate the anisotropies with position in the molecule is more difficult, however, and there are no simple rules to explain the variations in anisotropy along the sequence.

The foregoing discussion on amplitudes and anisotropies emphasizes the similarities between the results for nucleic acids and proteins (section 5.2.1). It should be pointed out, however, that the atomic motions are much more anisotropic in tRNA than in proteins. For example, the average ratio of the atomic mean square displacements along the axes of smallest and largest displacement is 0.11 in $tRNA^{Phe}$ (Prabhakaran *et al.*, 1985*a*) versus 0.26 in cytochrome c (Northrup *et al.*, 1981). Also, examination of the higher order moments of the atomic position probability distribution functions (the skewness and kurtosis) reveals that motions in tRNA are much more anharmonic than in proteins. Since the tRNA simulation is the only one for which the anisotropy and anharmonicity have been analyzed in detail, and since a complete discussion is available elsewhere (Prabhakaran *et al.*, 1985*a*), only a brief summary is given here.

There are several differences between the anisotropies in tRNA and in proteins. First is the remarkable difference in the average anisotropies, as described above. Second, there is a wider range of anisotropy ratios in tRNA, a phenomenon which is generally still true when we examine variations in anisotropy within different groups of atoms. Third, there is a striking difference in the dependence of anisotropy on accessible surface area: anisotropies are more pronounced in the interior of tRNA, while the opposite is true for the protein. These differences are almost certainly a consequence of the very different nature of nucleic acid structure as compared to protein structure. Crudely speaking, the double helical regions of tRNA are made up of stacked bases with the ribose–phosphate backbone wrapped around the outside. The stacked bases find themselves in a highly structured environment, with strong forces opposing their motions along the axis of the helix and weaker forces opposing sliding motions. The anisotropic nature of these forces produces very anisotropic motions.

Anharmonicities in the potentials of mean force are manifested by deviations of the atomic probability distribution functions from three dimensional Gaussians; cf. equation (5.1). In general, these anharmonicities are more pronounced in tRNA than in proteins, as reflected both by average values and the ranges of observed values of skewness and kurtosis. Again in contrast with the protein case, the kurtoses are more pronounced for atoms in the interior of the tRNA than for atoms on the surface. Another difference is that motions in tRNA along the eigenvector for the direction of smallest displacement show more skewed probability distribution functions than those along the eigenvector for the direction of greatest displacement; this situation is reversed for proteins. The reason for these differences in the anharmonicities of atomic motions of proteins and nucleic acids is not clear, although the explanation must be related to the structural factors responsible for the differences in anisotropies, as discussed in the previous paragraph.

5.3.2 *Hydrogen bond dynamics*

Certain common dynamic features of hydrogen bonds have been reported in some of the DNA simulations and in the tRNA simulation. First, there is a correlation between the average length of a hydrogen bond and the amplitude of the fluctuations in the bond length (Prabhakaran *et al.*, 1983; Singh *et al.*, 1985). In the case of tRNA, it is not surprising that secondary structure hydrogen bonds are both shorter and stiffer than tertiary structure hydrogen bonds, since the former consist of sets of two bonds (in A-U basepairs) or three bonds (in G-C basepairs), while many of the tertiary bonds are isolated single bonds. This pattern also reflects the relative stabilities of these bonds in the absence of magnesium (Crothers, 1979; Johnston & Redfield, 1979). The very high correlation with the class of bonds and the fact that a similar correlation is seen in Watson–Crick hydrogen bonds in DNA are very striking. In the case of tRNA, a further correlation is observed between the amplitude of fluctuations in the length of a hydrogen bond and the relaxation time for the decay of the autocorrelation function for that length. This is another reflection of the stiffness of the hydrogen bond, and this correlation is predicted for any Langevin oscillator (Berkowitz & McCammon, 1981). It has been observed for a variety of displacement correlation functions in proteins (Swaminathan *et al.*, 1982) and in tRNA (Prabhakaran *et al.*, 1983).

Second, when the correlations in lengths of pairs of hydrogen bonds are examined, an interesting pattern emerges. It would be expected that two hydrogen bonds in the same Watson–Crick basepair should show correlated motions, and such correlations are observed for bonds in A-U and A-T basepairs and for adjacent bonds in G-C basepairs (Levitt, 1983*d*; Harvey *et al.*, 1984; Singh *et al.*, 1985). In the case of the outer two bonds in the G-C basepair, small, generally negative correlation coefficients are observed (Harvey *et al.*, 1984; Singh *et al.*, 1985). Taken individually, these coefficients are not statistically significant, but the application of a simple sign test to the whole set of coefficients gives a highly statistically significant result in the tRNA case, since 11 of the 12 coefficients are negative ($P < 0.01$). Although the sample is too small to provide statistical significance in the DNA case, all four coefficients are negative. These results can be explained by the in-plane rocking of G-C basepairs, stretching one outside hydrogen bond while shortening the other.

Molecular dynamics simulations do not generally produce spontaneous breakage of basepairs inside otherwise intact double helices, so they do not provide direct information on the kinds of structures that may be responsible for proton exchange. These structures have been studied by other methods (section 6.4).

5.3.3 *Dynamics of backbone torsional angles*

As a general rule, most backbone torsion angles have values that fluctuate in a narrow range; the transitions between *trans*, *gauche*+ and *gauche*– configurations are rare. The amplitudes of these fluctuations have been reported to be in the range 6–12° by Levitt (1983*d*) and by Harvey *et al.* (1985*a*), while Singh *et al.* (1985) reported fluctuations over a range 9–21°. The source of these differences is not clear. For comparison, the root mean square fluctuations for the ϕ and ψ backbone torsions in proteins are 14–23° (van Gunsteren & Karplus, 1982*b*; Levitt, 1983*b*). There are substantial differences in the relative mobilities of the different backbone torsions for the tRNA and DNA simulations. The most interesting result concerns δ, the torsion within the ribose ring. In the tRNA simulation (Harvey *et al.*, 1985*a*), it is the most restricted torsion angle, while in the comparable DNA simulation, without explicit counterions (Singh *et al.*, 1985), it is the second most mobile torsion. It is possible that this difference reflects the dependence of δ on sugar pucker, because changes in pucker for C3′-*endo* sugars produce much smaller changes in δ than do variations in pucker for C2′-*endo* sugars. This leads to a pucker-dependent flexibility for δ, with C3′-*endo* sugars relatively more rigid than C2′-*endo* sugars (Westhof & Sundaralingam, 1983*a*,*b*).

A detailed examination of the size of the fluctuations in backbone torsions in Z-DNA has shown that the molecule is in dynamic equilibrium between several structures, including the crystallographically observed Z_I and Z_{II} forms (Wang *et al.*, 1979; Ramstein & Leng, 1980). In the molecular dynamics simulation of Westhof *et al.* (1986), it was observed that those torsions that showed the smallest rms fluctuations are those whose values do not change much on going from the Z_I to the Z_{II} configuration, whereas the torsions whose values have the largest differences between the Z_I and Z_{II} forms are the same ones that show the largest fluctuations in the molecular dynamics simulation.

There are a number of pairs of torsional angles whose motions are found to be correlated in the DNA simulations (Levitt, 1983*d*; Singh *et al.*, 1985). Among those pairs that are separated by one bond, negative correlation coefficients would be expected for crankshaft rotations, which should occur if the torsion for the intervening bond is in the *trans* configuration (Olson, 1981). An example of this is a crankshaft rotation about the O5′—C5′ bond, in which the pair of torsions (α, γ) goes from a (*gauche*−, *gauche*+) configuration to a (*trans*, *trans*) configuration. This motion has been suggested to occur in the B to Z transition (Olson, 1981; Sundaralingam & Westhof, 1981; Harvey, 1983) and as part of the longitudinal stretching that separates basepairs in the formation of intercalation sites (Takusagawa & Berman, 1983; Daune *et al.*, 1985). Even in the absence of full transitions from the *gauche* to *trans* configurations, the dynamic fluctuations in α and γ do show the expected negative correlation coefficient (Singh *et al.*, 1985). In addition, since ε is in the *trans* configuration in B DNA, a similar result would be predicted for the pair (δ, ζ), and this negative correlation is also observed. There is one positive correlation coefficient of note, for the pair (β, δ), but the intervening torsional angle, γ, is *gauche*+ rather than *trans*, so no crankshaft motions would be expected. The crankshaft correlation (α, γ) is also reported in a statistical analysis of the correlations of pairs of static values of torsional angles in a $tRNA^{Phe}$ crystal structure (Kitamura, Mizuno, Amisaki, Tomita & Babas, 1984). Other pairs of angles whose motions are predicted to be correlated in that statistical analysis are (α, β), (β, ε), and (ε, ζ), and these are also among the predominant correlations observed in the DNA molecular dynamics simulation of Singh *et al.* (1985).

5.3.4 *Dynamics of double helical regions*

Although the fiber diffraction studies on DNA structure led to double helical models where all basepairs had identical geometries (Watson & Crick, 1953; Arnott *et al.*, 1976), it is now clear that the double helix has local variations that depend on sequence. These effects have been observed both crystallographically and in solution studies, and they have led to a number of theoretical studies aimed at understanding the rules that govern the sequence-dependence of the local geometry of the double helix (Calladine, 1982; Fratini *et al.*, 1982; Kabsch, Sander & Trifonov, 1982;

Dickerson, 1983; Haran, Berkovich-Yellin & Shakked, 1984; Tung & Harvey, 1984, 1986*a*,*b*).

The dynamics of double helical regions of nucleic acids can be expected to depend strongly on overall conformation, being different for RNA and A-, B-, and Z-DNA. Within a given structure, they should also show some dependence on sequence.

The most extensive analysis of the dynamic aspects of helix structure is the normal mode analysis of three DNA hexamers by Tidor *et al.* (1983). They used the same potential function for a 60 ps molecular dynamics simulation on $d(CGCGCG)_2$ in the B conformation and for a normal mode analysis of this same structure. Obtaining similar results for a variety of motions in these two studies, they concluded that the harmonic approximation provides a reasonable approach for investigating motions in DNA at room temperature. For comparative purposes, they carried out normal mode analyses on the same hexamer in the Z conformation and on $d(TATATA)_2$ in the B conformation.

The model of Tidor *et al.* (1983) did not include explicit hydrogen atoms except for those that can form hydrogen bonds. Consequently, all but a handful of the 816 normal modes of the B form $d(CGCGCG)_2$ have frequencies below 2000 cm^{-1}. The fact that the mean square amplitude of the atomic motion is inversely proportional to the frequency (for similar effective masses) causes the 50 lowest frequency modes to account for over 80% of the motion. These frequencies all lie below 100 cm^{-1} ($\tau = 0.3$ ps), and the lowest frequency mode for all three analyses is near 10 cm^{-1} ($\tau = 3$ ps). Because these low frequency normal mode motions are collective motions, and because they do account for most of the motion of all the atoms, they will reflect the time scale of fluctuations in the local geometry of the double helix. Given the statistical sampling problem (Zwanzig & Ailawadi, 1969), molecular dynamics simulations would have to cover several tens of ps to examine correlations and characteristic times for these motions.

That the modes in the 10 cm^{-1}–100 cm^{-1} range really do account for helix dynamics is most vividly seen in movies of these modes. Among the motions reported by Tidor *et al.* (1983), after examining such movies, were bending and twisting of the helix, longitudinal oscillations with successive basepairs moving out of phase, base sliding, basepair buckling, propeller twisting, and a variety of backbone modes. With regard to the latter, they observed that the major groove was deformed more than the minor groove. They also found that the backbone motions for cytidine and guanosine were very similar for the B form alternating C-G hexamer, whereas the thymidine and adenosine backbones have very different motions in the B

Table 5.2. *Torsional stiffness of B-DNA*

Method	Torsional modulus (10^{-19} erg cm^{-1})	References
Theoretical:		
Normal mode analysis	13	Tidor *et al.* (1983)
Conformational energy calculations	5.9	Tung & Harvey (1984, 1986*b*)
Experimental:		
Electron paramagnetic resonance	3.1	Robinson *et al.* (1980)
Cyclization probability	2.4	Shore & Baldwin (1983)
Supercoiling	0.6–4.4	Depew & Wang (1975); Pulleyblank *et al.* (1975); Vologodskii *et al.* (1979); Strogatz (1982)
Fluorescence depolarization	0.4–4.0	LeBret (1978); Barkley & Zimm (1979); Thomas *et al.* (1980); Millar *et al.* (1981)
Phosphorescence depolarization	0.4–14.0	Hogan *et al.* (1983)

form alternating A-T hexamer. In the Z oligomer, the C and G backbones also have very different motions, but this is to be expected in light of their very different average (crystallographic) structures. Overall, the Z-DNA structure was found to be stiffer than the B-DNA structure, which agrees with experiment (Wang *et al.*, 1979; Ramstein & Leng, 1980; Mirau, Behling & Kearns, 1985).

The torsional stiffness of DNA has been measured by a variety of methods (Table 5.2). The mean square fluctuation in helix twist angle between adjacent basepairs is

$$\langle \theta^2 \rangle = ak_B T/C,$$

where a is the distance between the basepairs, $k_B T$ is the thermal energy, and C is the torsional modulus. The root mean square amplitude of the helix twist angle in the normal mode analysis of Tidor *et al.* (1983) was 2.0°, about half the experimental value. A simple model of helix twisting that uses equivalent torsional springs whose torque constants are determined by an extensive grid search of the conformational space of the bases gives an amplitude of 3° for $d(CGCGCG)_2$ in the B conformation (Tung & Harvey, 1986*b*). A comparison of the experimental values of the torsional stiffness of B-DNA with these theoretical values is given in table 5.2. Both the normal mode analysis (Tidor *et al.*, 1983) and the torsional equivalent spring model (Tung & Harvey, 1986*b*) predict that helix twisting motions of adjacent basepairs will be anticorrelated.

Experimentally, perhaps the most extensive analysis of motions in DNA is that of Holbrook & Kim (1984). They determined the directions and amplitudes of motions of phosphate, ribose, and base moeities using a 'segmented rigid body' analysis of X-ray crystallographic data. The motions they reported included propeller twisting and basepair rolling, buckling and sliding of basepairs, coupled rotations of sugars and bases as a unit, coupled translations of basepairs, helix bending, and fluctuations in the widths of the major and minor grooves.

5.4 Nature of short time dynamics

From the studies summarized in this chapter, a fairly consistent picture emerges for atomic motions in proteins and nucleic acids. Individual atoms have vibrational motions that are rapid and of small amplitude (hundredths of a nm, hundredths of a ps), superimposed on the slower, larger amplitude motions of groups of atoms (motions on the order of 0.1 nm with characteristic times of about 10 ps). The differences in amplitude of the atomic displacements in different parts of a protein or nucleic acid are primarily due to the slower, collective motions. Frequent collisions render the atomic motion chaotic, so group motions on the subpicosecond scale are diffusive, like those in a liquid. Over longer time scales, as groups of atoms wander away from their average, crystallographically observed positions, significant restoring forces come into play. The group motion is then oscillatory about the average position, but with significant damping from collisional effects. Since the magnitude and direction of atomic motions are dominated by group displacements about their mean positions on the time scale of 10 ps, atomic motion on this time scale has a solid-like character, particularly in the interior of globular proteins. This qualitative picture of the hybrid solid/liquid character of macromolecular structure, with the solid character most pronounced in the interior, is consistent with a simple theoretical model of protein structure developed some time ago by Lifshitz (1969). Even though many of the modeling studies that produced this picture were done *in vacuo*, the inclusion of explicit solvent does not fundamentally change this view.

The nature of the collective motions that dominate the atomic displacements in proteins and nucleic acids is still a subject of active investigation. Because normal mode calculations account for a sizable fraction of the total atomic displacement observed in molecular dynamics and X-ray diffraction studies, thermally excited elastic distortions about a mean conformation must be an important component of the collective motion. *In vacuo*, these motions would have periods up to 10 ps or so. The actual motion will have a more complex character due to the superposition of different modes, anharmonic coupling and, for molecules in solution, solvent damping. In addition to these elastic motions, there are also conformational transitions corresponding to motions of the molecule among different local minima in the neighborhood of the average structure on the potential energy surface. Such transitions are commonly observed in molecular dynamics simulations. Also, kinetic studies of ligand binding to heme proteins show that these proteins can be frozen into different substates at sufficiently low temperatures (Austin *et al.*, 1975; Ansari *et al.*, 1985). Since these transitions involve the crossing of energy barriers, the times required to sample a fully representative set of substates at room temperature may be of the order of 10^{-7} s (Ansari *et al.*, 1985). Such substates may be analogous to the defective lattice configurations or 'hidden structures' that are obtained by using energy minimization to relax configurations of liquids to the nearest local energy minima (Stillinger & Weber, 1984). From this point of view, a protein, for example, differs from a liquid primarily as a result of the constraints imposed by the covalent connectivity of the polypeptide chain. These constraints broaden the spectrum of structural relaxation times for the protein by increasing the energy barriers for larger conformational transitions, and introduce strong couplings between local and collective displacements (McCammon, 1984; Ansari *et al.*, 1985; Stein, 1985).

The large contribution of collective motions to the net atomic displacements suggests that such motions are of importance in macromolecular function. This supposition is supported by studies of activated processes, in which collective motions are found to play an essential role, and by studies of the large scale motions that are involved in ligand binding and macromolecular interactions. Those motions are discussed in more detail in the next two chapters.

6

Local structural transitions

6.1 Introduction

Local structural transitions are of great importance in biological activity. The binding and release of ligands by proteins or nucleic acids often involves displacements of certain groups within the macromolecule as well as the displacement of the ligand itself. The rates of some enzyme–substrate reactions are determined by the time required for rearrangement of covalent bonds in elementary reaction steps, or by the time required for reorientation of a catalytically important sidechain between such steps. Also, important changes in the large scale structure of biopolymers may sometimes proceed *via* a sequence of more localized transitions.

Local transitions generally occur on relatively long time scales because they are activated processes. That is, they involve the crossing of free energy barriers that separate the initial and final configurations. Such transitions are occasionally observed during conventional molecular dynamics simulations. For example, three backbone amide groups underwent 180° rotations in the first protein simulation (McCammon *et al.*, 1977). Examination of these revealed that structural relaxation and collisional damping due to the motion of neighboring protein atoms determine the character of the local transitions. Because such transitions occur infrequently, they cannot be studied systematically by the conventional simulation approach, however. The adiabatic mapping method described in section 4.5 provides a systematic but approximate method for studying local transitions. The activated dynamics method described in section 4.9 provides a detailed approach, but is more demanding of computational resources.

6.2 Results for proteins

Theoretical studies of local structural transitions in proteins show that such transitions often have a partly collective character. As has been discussed in section 5.2.2, this result is expected in view of the dense packing of atoms in proteins and their solvent surroundings. The displacement of any atom through a distance comparable to its radius requires concomitant displacements of the neighboring atoms. Typical results of this kind have been obtained in studies of ligand binding by myoglobin. It has long been recognized that structural fluctuations must occur for oxygen or other ligands to be able to move between the densely packed segments of polypeptide that separate the myoglobin surface from the heme binding site in the interior of the protein (Perutz & Mathews, 1966). No static channels of sufficient width are apparent in the X-ray structures. Case & Karplus (1979) identified two potential pathways for ligand insertion and applied empirical energy functions to estimate the energetic cost of widening these pathways sufficiently to allow ligand insertion. These estimates were obtained by fixing a model ligand at local bottlenecks on each path and then allowing the protein to relax by energy minimization. The ligand was represented by a sphere with an effective radius of 0.32 nm; the local bottlenecks were between the distal His E7 and Val E11 of the E helix and between His E7 and Thr E10 on the first path, and between Phe B14 and Leu E4 on the second path. It was found that the energy barriers to ligand penetration, which would exceed 400 kJ/mol if the structure were rigid, drop to values of 54 kJ/mol or less on the first path and 75 kJ/mol on the second path after energy minimization. The protein structural changes upon relaxation include shifts in the sidechains indicated above and, in addition, smaller shifts in neighboring sidechains and in the backbones of helices D and E. These shifts are thought to resemble some of the spontaneous structural fluctuations that facilitate ligand insertion in the true dynamic situation. As discussed in section 4.5, adiabatic mapping calculations often overestimate the enthalpy barriers for local conformational transitions. Indeed, recent dynamical relaxation studies suggest that the free energy barrier for ligand escape from the heme pocket in myoglobin may be dominated by entropy rather than enthalpy at 300 K. These studies are described in chapter 9.

More complete pictures of the dynamic aspects of local structural transitions can be obtained by the activated dynamics simulation method described in section 4.9. This method has been applied in a detailed study of rotational isomerization of tyrosine sidechains in the interior of the protein BPTI (Northrup *et al.*, 1982*a*; McCammon *et al.*, 1983). In this process, the tyrosine ring rotates such that $\Delta\chi^2 = 180°$ (cf. figure 2.3). The

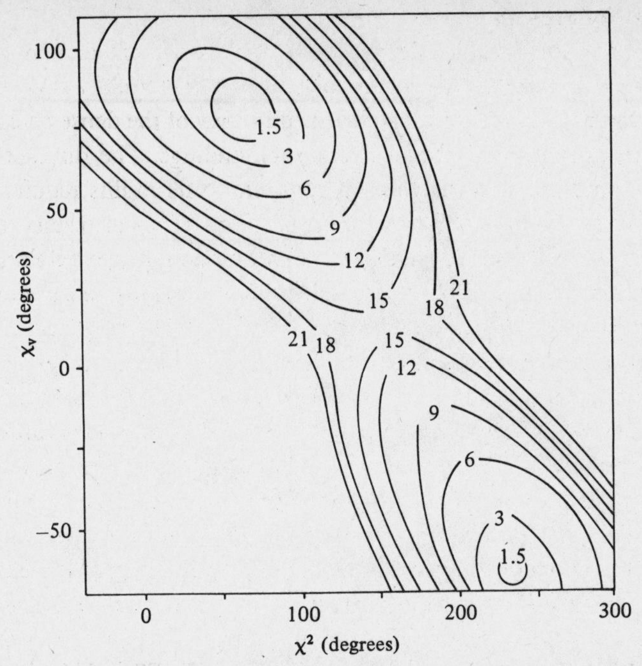

Fig. 6.1. Adiabatic potential surface for rotational isomerization of the tyrosine 35 ring in the pancreatic trypsin inhibitor (McCammon *et al.*, 1983). The contours are in units of kcal/mol (1 kcal = 4.184 kJ). (Copyright © 1983, American Chemical Society. Reprinted with permission.)

energy barrier to this rotation arises from nonbonded repulsions between atoms in the ring and in the surrounding protein matrix. An early analysis, using energy minimization methods, showed that the very large barriers encountered in a hypothetical rigid protein are reduced to values in the range 40–100 kJ/mol when deformations of the matrix are allowed (Gelin & Karplus, 1975). Although this is a comparatively simple activated process, the full dynamical analysis has provided a number of important general results (Northrup *et al.*, 1982*a*; McCammon *et al.*, 1983). A significant initial finding was that the dihedral angle χ^2 is not by itself a good reaction coordinate. Examination of structural models showed that, during rotation of the tyrosine 35 ring, certain ring atoms come into close contact with atoms of the adjacent polypeptide backbone, particularly with N_{36}. Thus, a new reaction coordinate was constructed as $\xi = \chi^2 - \chi_V$; χ_V is an angle that measures how far N_{36} is from the plane of the ring. Energy minimization of the system was carried out for fixed values of χ^2 and χ_V. The resulting adiabatic map provides a qualitative indication of the effective potential surface for the motion of these two

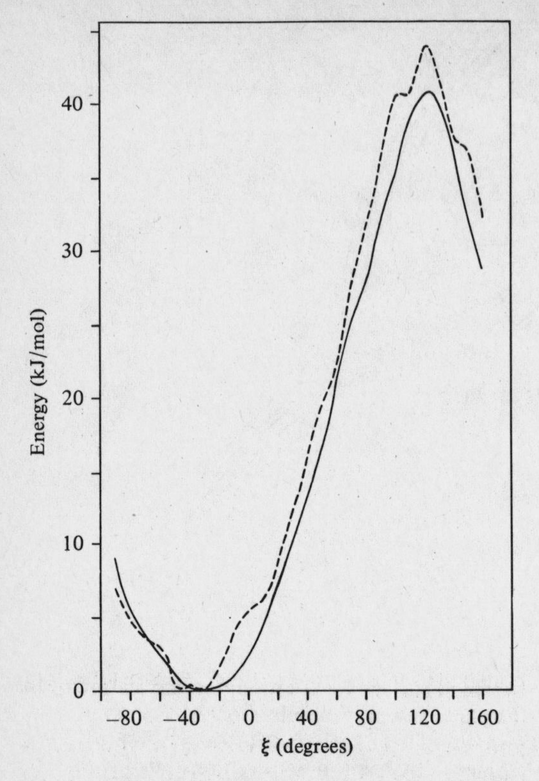

Fig. 6.2. The potential of mean force $A(\xi)$ (——) and mean potential energy $\langle V(\xi)\rangle$ (- - -) as functions of the tyrosine 35 ring rotational isomerization reaction coordinate ξ (Northrup *et al.*, 1982*a*). The energies are given in units of kcal/mol (1 kcal = 4.184 kJ).

degrees of freedom. From the shape of this potential surface (figure 6.1) the reaction coordinate is seen to provide a good description of the most likely path of motion from the initial state to the final state of the system. The umbrella sampling calculations carried out using this reaction coordinate yield the results shown in figure 6.2. The potential of mean force $A(\xi)$ is related to the probability $\rho(\xi)$ of occurrence of a fluctuation with the value ξ by the equation

$$A(\xi) = -k_B T \ln \rho(\xi) \tag{6.1}$$

Here, A describes the work function or Helmholtz free energy (assuming temperature and volume are fixed) of the system as a function of the reaction coordinate. The second curve in the picture describes the variation of the average potential energy of the system as a function of the reaction coordinate. From the relation $\Delta A = \Delta E - T\Delta S$, and the fact that

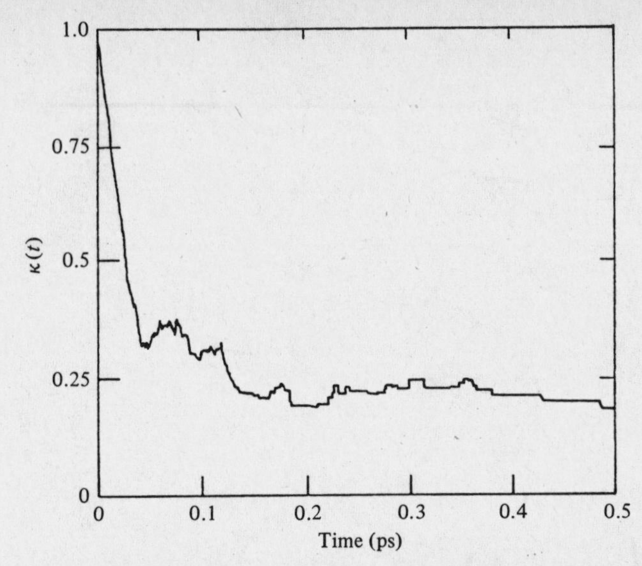

Fig. 6.3. The normalized reactive flux time correlation function $\kappa(t)$ for the tyrosine 35 ring rotational isomerization (Northrup *et al.*, 1982*a*).

$\Delta E = \Delta\langle V\rangle$ when T is constant, it is seen that the intrinsic entropy variation during the rotation is small.

The calculation of trajectories starting from the barrier region showed that a substantial fraction of these did not cross the barrier in the smooth, uninterrupted manner assumed in the 'ideal transition state theory' (see section 6.4). Frictional effects, particularly collisions between the ring and the local backbone, were found to interrupt a significant number of the crossing attempts. The transmission coefficient κ for the process (cf. section 4.9) was found to be about 0.2 (figure 6.3). Detailed analysis of the individual trajectories has helped considerably to clarify the mechanism for this isomerization process (McCammon & Karplus, 1980*b*; McCammon *et al.*, 1983). The ring rotation was found to be preceded by important structural fluctuations in the matrix around the ring. Particularly significant is a spontaneous displacement of a section of backbone that lies above one face of the ring. This displacement systematically precedes the ring rotation and appears to make two important contributions. The first effect of the backbone fluctuation is to displace atoms slightly from the path of the ring, resulting in a substantial reduction of the barrier to be crossed. Thus, the ring rotation appears to be gated by a collective motion in the surrounding matrix. During the rotation, bond angle deformations occur within the sidechains and adjacent backbone; the effect of these is to increase the distance of closest approach between

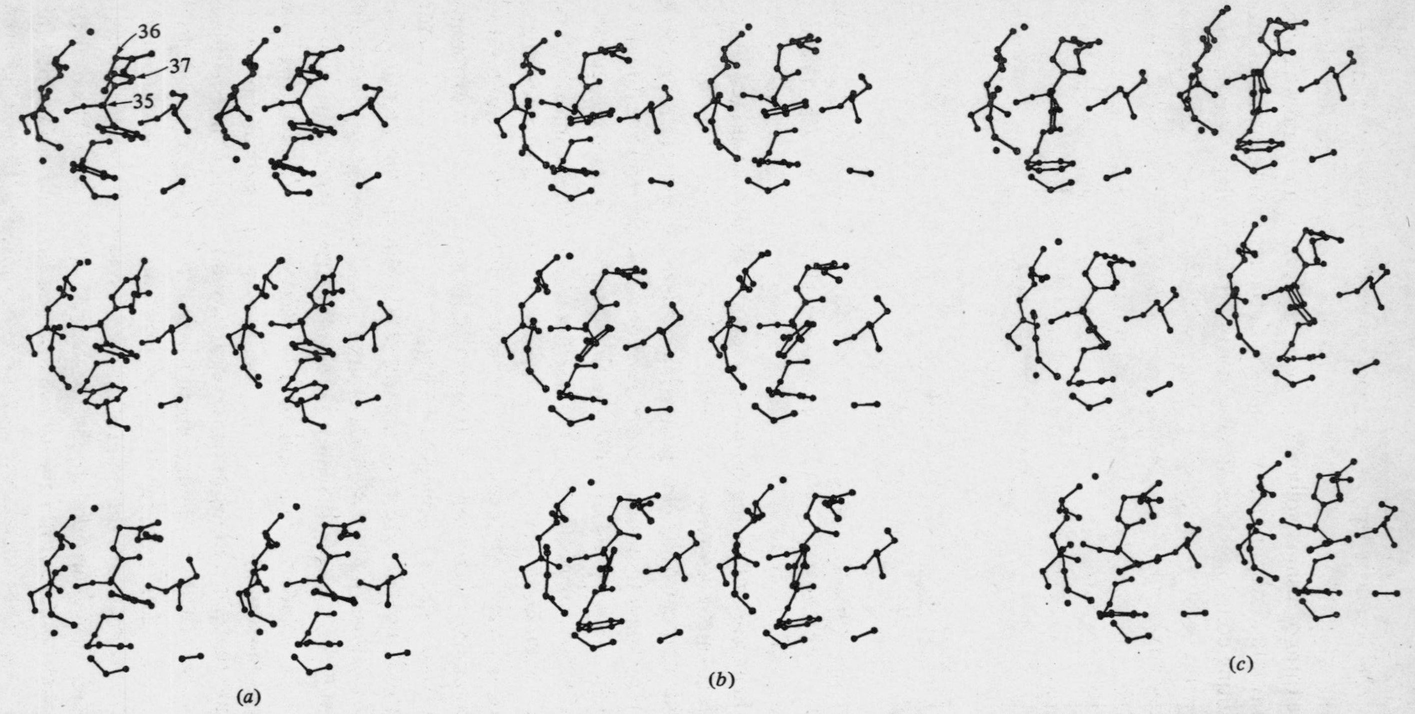

Fig. 6.4. Stereoviews of atoms in the vicinity of tyrosine 35 at a sequence of times during rotational isomerization (McCammon *et al.*, 1983). From top to bottom in each panel, the times in ps are (*a*) 0.50, 0.75, 1.00 (*b*) 1.25, 1.38, 1.50 (*c*) 1.62, 1.75, 2.00. The gate, which includes the backbone of residue 37, opens about 0.5 ps before the tyrosine 35 ring begins to rotate. (Copyright © 1983, American Chemical Society. Reprinted with permission.)

the ring and N_{36}, resulting in a further reduction of the effective energy barrier. The second effect of the backbone fluctuation is to create a transient packing defect that helps to initiate the transition. The defect takes the form of a small volume (approximately 0.01 nm^3) into which the ring can rotate in response to collisions with the remaining matrix atoms. The resulting collisional bias allows the ring to accumulate the rotational kinetic energy necessary to surmount the residual barrier and is reminiscent of the mechanism of displacement of atoms in simple liquids (Rahman, 1966). The coupling of the collective fluctuation in the protein matrix and the local isomerization reaction is apparent in a sequence of structural snapshots, figure 6.4.

A number of experimental data are available for comparison with theoretical results such as those described above. As in the case of the short time scale motions, there is qualitative agreement even where there are differences of detail. The rebinding of ligands to myoglobin and related proteins following flash photolysis has been studied under a wide variety of conditions by Frauenfelder and coworkers (Austin *et al.*, 1975; Beece *et al.*, 1980; Debrunner & Frauenfelder, 1982; Ansari *et al.*, 1985). These studies suggest that a ligand must cross energy barriers in the course of moving from solvent to the heme binding site. While the origin of the barriers is still not known in detail, some appear to be associated with steric constraints presented by the polypeptide chain. The heights of these barriers apparently fluctuate rapidly at physiologic temperature (see section 6.4), but are certainly much smaller than the 400 kJ/mol estimated for rigid proteins. Moreover, the rebinding kinetics depend on solvent viscosity; this suggests that the protein fluctuations that facilitate ligand displacement are partly collective in character. An X-ray diffraction study of myoglobin with a phenyl group bound to the heme iron has recently been described by Ringe, Petsko, Kerr & Ortiz de Montellano (1984). This comparatively large ligand effectively holds open a channel between the iron and the protein surface. Although the changes in protein structure may be more dramatic than those required for entry of a small ligand (e.g., changes of 100° in the χ^1 dihedral angles of His E7 and Val E11, and the shuffling of ion pair interactions involving several sidechains), the results do indicate that some collective character is likely to be involved in the opening of gates even for small ligands.

The kinetics of tyrosine and phenylalanine ring rotations in a number of proteins have been studied by NMR. At low temperatures, the rings may remain for a sufficiently long time in given orientations that the two δ and two ε protons can be individually distinguished by separate spectral lines because of their different magnetic environments. The rings rotate more rapidly as the temperature is increased, producing spectral changes that lead eventually to coalescence of some of the lines when rapid rotation produces an equivalent average magnetic environment for the corresponding protons. Analysis of these spectral changes allows determination of the rate constants, activation energies and activation entropies for the ring rotations. For the tyrosine 35 ring in BPTI, the experimental free energy barrier to rotation (Wagner, DeMarco & Wüthrich, 1976) is 30% larger than the theoretical results described above. This quantitative discrepancy appears to be associated with the omission of solvent water from the initial calculations (Northrup *et al.*, 1982*a*). In the absence of hydrogen bonding and other interactions with the solvent, the surface of the protein (which includes part of the ring environment) changes somewhat from the expected solution structure; the slight distortions that result in the protein matrix are sufficient to account for the discrepancy.

6.3 Results for nucleic acids

Many investigators have done static modeling studies on nucleic acid structures, using either physical or computer-based models, and some of these studies have suggested possible pathways for local structural changes. The interested reader will find representative publications from most of the research groups in the books edited by Sarma (1981*a*, *b*), Clementi & Sarma (1983), and Rein (1985), and in volume 47 of the Cold Spring Harbor Symposium on Quantitative Biology (1983). We will confine ourselves to a brief survey. We first describe a few of the many studies carried out by Kollman and his coworkers, in which energy minimization methods have been used to examine structures and energetics of a variety of local conformational changes in DNA; a more complete review of this work has been given by Kollman *et al.* (1985). We will consider only those results with explicitly dynamic consequences. We then briefly describe the repuckering of ribose and deoxyribose rings based on examples of spontaneous repuckering observed in molecular dynamics simulations. Third, we discuss local transitions accompanying the formation of intercalation sites. Finally, we report on the configurational entropy difference between B-DNA and Z-DNA.

Energy minimization can be used to investigate the energetics and possible pathways of local conformational changes if those changes can be forced to occur in a model molecule. Shortly after the crystal structure of a B-DNA dodecamer (Wing *et al.*, 1980) revealed a number of departures from the ideal double helical geometry of the classical model of DNA structure, the question of the relationship between the local helix twist angle and the backbone conformation was examined by Kollman, Keepers & Weiner (1982). They built a series of model oligomers that were identical in every respect except for the helix twist angle between successive basepairs, and they refined these with energy minimization. Because of the limited structural changes produced by the minimization procedure, they were able to identify a number of local features whose changes were coupled to helix twisting motions. In particular, they observed that with a helix repeat of nine residues per turn, all sugar puckers remained in the C2′-*endo* configuration on refinement, whereas with underwound helices with a repeat of 13 residues per turn, there was a pronounced tendency of sugars to repucker to the O4′-*endo* and C3′-*endo* configurations. These results are, of course, consistent with the well-known relationship between sugar pucker and helix twist observed in fiber diffraction (Arnott *et al.*, 1976). With their most carefully refined model, which has an average helix repeat of ten residues per turn, Kollman *et al.* (1982) found that pyrimidines were more likely to repucker than purines, in agreement with the crystallographic results (Drew *et al.*, 1981).

In other studies, Keepers, Kollman, Weiner & James (1982) and Keepers, Kollman & James (1984) examined the kinds of local conformational changes that might be expected to accompany basepair melting. After refining normal double helical models of DNA, they introduced an additional constraint term into their potential function, an interaction between pyrimidine N3 and purine N1 atoms that would force the hydrogen bond to stretch from its normal length (about 0.3 nm) to a length of 0.6 nm, effectively rupturing the bond. During additional energy minimization, the models would move in response to this forced hydrogen bond melting, presumably along pathways that could provide information on the mechanism of proton exchange. When the constraint was later removed and additional energy minimization was done, some of the structures reverted to the original Watson–Crick form, while others relaxed to local low energy conformations in which the imino proton would be accessible to the solvent. Some of these structures involved the formation of a new hydrogen bond between the thymine O2 and the adenine amino proton at N6. In other structures, the two bases had slid past one another with the pyrimidine projecting into the major groove and the purine projecting into the minor groove. Although appropriately cautious about over-interpreting the energies calculated by their procedure, Keepers *et al.* (1982) did note that the energies were consistent with the enthalpy changes observed in the hydrogen exchange experiments of Mandal, Kallenbach & Englander (1979), suggesting that the models of melted basepairs were good candidates for the true structures. As expected, the principal energetic costs of this model melting were those associated with the rupture of the Watson–Crick hydrogen bonds and with the loss of base stacking. In their later study (Keepers *et al.*, 1984), the effects of sequence and conformation were examined in detail, producing a number of results that agreed with experimental studies on proton exchange, and providing information on the sequence dependence of the backbone motions and sugar puckering changes that accompany basepair opening. In particular, their results are consistent with proton exchange experiments that indicate that the rate of exchange for A-T basepairs is relatively independent of the surrounding base sequence (Mirau & Kearns, 1983). It is also worth noting that NMR relaxation studies suggest that the *kinds* of motions studied by Keepers *et al.* (1982, 1984) do play a role in proton exchange when such motions occur on the nanosecond time scale (Mirau *et al.*, 1985).

The dynamics of sugar puckering have been studied in detail as a consequence of the spontaneous repuckering of several sugars in molecular dynamics simulations of DNA (Singh *et al.*, 1985) and of tRNA (Harvey *et al.*, 1985*a*; Harvey & Prabhakaran, 1986). In all cases, passage between the northern and southern quadrants (i.e., C3′-*endo* and C2′-*endo* configurations) is via the eastern (O4′-*endo*) pathway, as expected. Repuckering is another example of a rare but intrinsically rapid event, with spontaneous repuckering for both riboses and deoxyriboses in single stranded or loop regions or at the ends of a double helix occurring with an average rate of about 1 ns^{-1} (Harvey & Prabhakaran, 1986). Repuckering is rapid, requiring only a few ps, as is seen in figure 6.5. The kinetics of the repuckering of deoxyriboses are very similar (Singh *et al.*, 1985). Figure 6.6 is a scatter diagram of 4800 representative ribose configurations from the tRNA simulation, presented on the traditional polar plot. This diagram shows that puckering is well described by the pseudorotation pathway, with a band of nearly constant puckering amplitude (Altona & Sundaralingam, 1972; Cremer & Pople, 1975). The average amplitude is 0.040 nm for riboses in the tRNA simulation (Harvey & Prabhakaran, 1986) and 0.035 nm for deoxyriboses in the DNA simulation (Singh *et al.*, 1985). It is not clear if this difference reflects a true difference in puckering behavior in riboses and deoxyriboses or if it is simply an artifact of differences in the potential functions. The deviations from the constant amplitude of true pseudorotation, with a larger amplitude at the O4′-*endo* configuration (figure 6.6), were first predicted by Levitt & Warshel (1978).

Regarding the relative ease with which purines and pyrimidines can repucker, it is interesting that the only two sugars that pass from C2′-*endo*

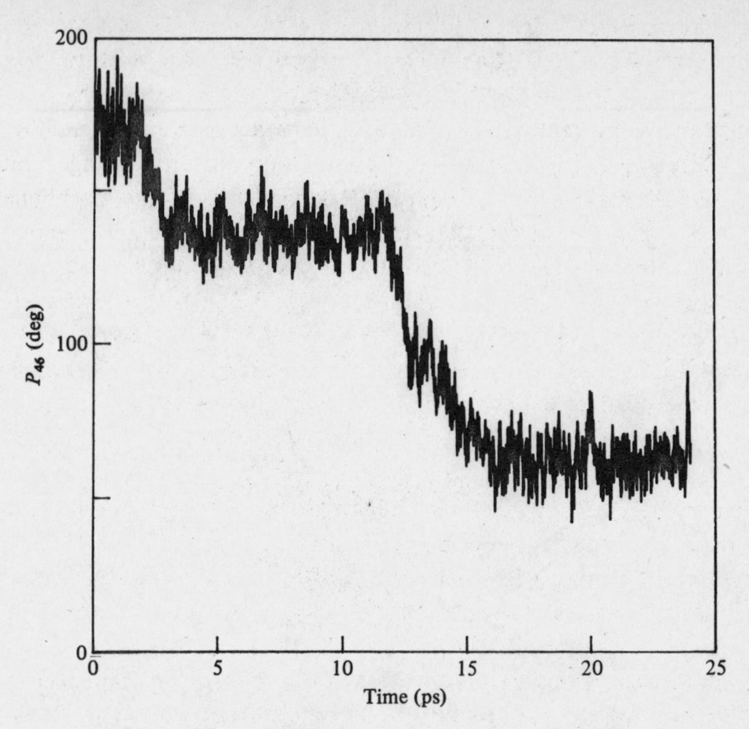

Fig. 6.5. Time course of the pucker angle for ribose 46 of $tRNA^{Phe}$ from a molecular dynamics simulation (Harvey *et al.* 1985*a*). There is a spontaneous repuckering from C2′-*endo*-like configurations ($P > 90°$) to C3′-*endo*-like configurations ($P < 90°$) that occurs in about 5 ps. (Copyright © 1985, John Wiley & Sons, Inc. Reprinted by permission.)

all the way to C3′-*endo* in the DNA simulation are pyrimidines (Singh *et al.*, 1985), as predicted from energy minimization studies (Kollman *et al.*, 1982) and from the crystallographic results (Drew *et al.*, 1981). On the other hand, the spontaneous repuckering in tRNA happens to two purines, m_2G46 and A76 (Harvey *et al.*, 1985*a*). In the Z-DNA molecular dynamics simulation (Westhof *et al.*, 1985) the purine sugars show a much broader range of fluctuations in pucker than do the pyrimidines. It is not clear what importance, if any, should be attached to these results, since the number of spontaneous repuckerings is much too small to provide any statistical significance. Molecular mechanics studies on alternating d(A-T) sequences have shown that C3′-*endo* conformations are favored for deoxyriboses on the 5′ side of thymines (Rao & Kollman, 1985).

There are a number of parameters of sugar conformation that are highly

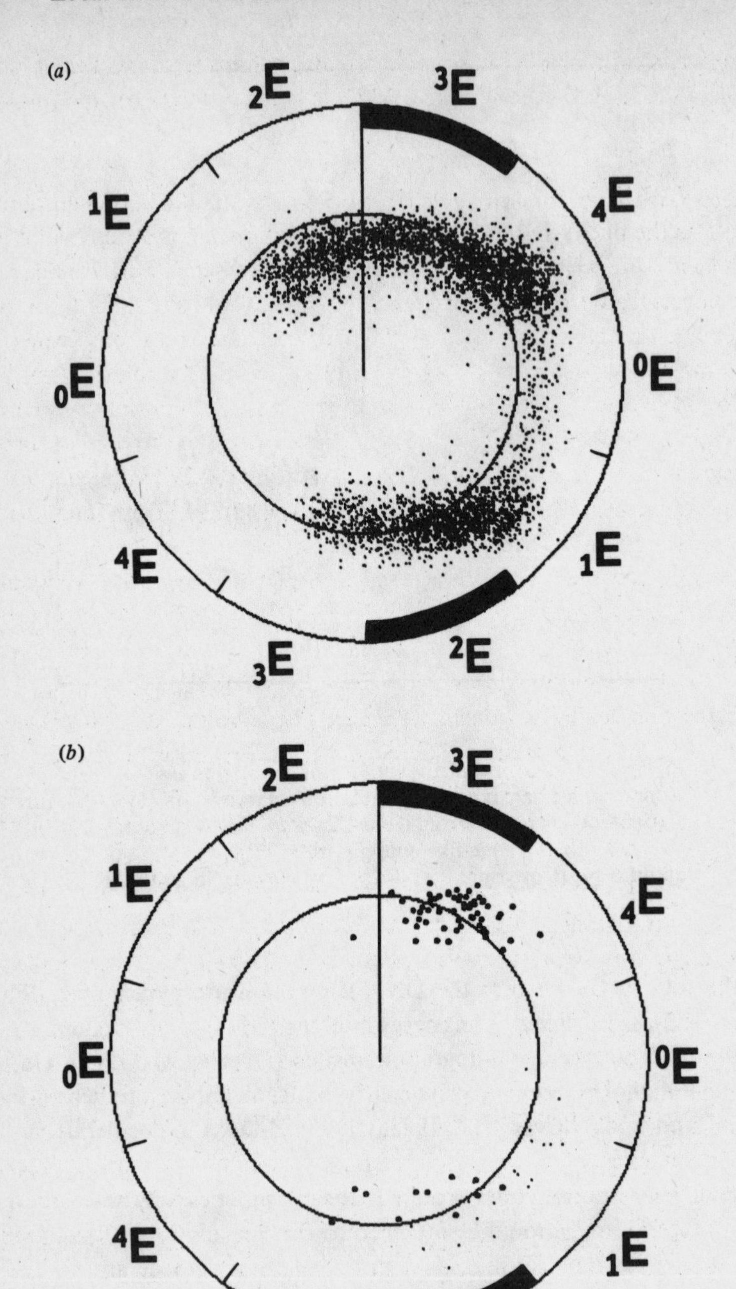
(a)
2E
3E
1E
4E
0E
0E
4E
1E
3E
2E
(b)
2E
3E
1E
4E
0E
0E
4E
1E
3E
2E

correlated with sugar pucker. The original definition of puckering phase angle (Altona & Sundaralingam, 1972) is given in terms of ring torsions:

$$\theta_j = \theta_m \cos (P + 4\pi j/5) \tag{6.2}$$

where θ_j is the jth torsion angle ($j = 0, 4$), θ_m is the puckering amplitude, and P is the puckering phase angle. This relationship is strictly valid only for ideal ring geometries, and Levitt & Warshel (1978) found small deviations from this behavior in their analysis of ribose flexibility. A more extensive analysis, based on 4800 ribose configurations generated spontaneously during the tRNA molecular dynamics simulation, has been reported (Harvey & Prabhakaran, 1986). Since the exocyclic torsion angle δ, defined by C5′-C4′-C3′-O3′, is so closely related to the endocyclic torsion θ_1, defined by O4′-C4′-C3′-C2′, δ would also be expected to follow a periodic relationship like (6.2). Data from the 4800 ribose configurations of the tRNA simulation are shown in figure 6.7, which shows small deviations from the expected behavior. A plot of θ_1 versus P (Harvey & Prabhakaran, 1986) is very similar to figure 6.7. Noting that δ is more sensitive to changes in P for C2′-*endo* sugars than for C3′-*endo* sugars, Westhof & Sundaralingam (1983*a*, *b*) argued that there is a dependence of backbone flexibility on sugar pucker. Their argument is illustrated by figure 6.7, where it is seen that partial pseudorotation in the C3′-*endo* domain leads to smaller changes in δ than the same change in P in the C2′-*endo* domain (solid line). Note, however, that the very large thermal fluctuations in the value of δ at any given P are sufficient to seriously diminish any effects of pucker-dependent backbone flexibility. Energy minimization studies also predict small but systematic variations of the bond angles within the furanose ring as it is moved along the pseudorotation

---

Fig. 6.6. (*a*) Polar scatter plot of the sugar pucker phase angle (angular coordinate) and puckering amplitude (radial coordinate) for 4800 ribose conformations generated during a $tRNA^{Phe}$ molecular dynamics simulation (Harvey & Prabhakaran, 1986). The outer circle corresponds to an amplitude $\theta_m = 70°$, while the inner circle corresponds to the average amplitude, 41.8°. The plot is divided into segments that cover the ten envelope configurations, indicated by the usual notation: ^2E = C2′-*endo*; $_2E$ = C2′-*exo*; ^0E = O4′-*endo*; etc. (*b*) To demonstrate that the scatter in figure 6.6(*a*) is not an artifact of the simulation, this diagram presents the same plot for the 76 crystallographic ribose structures in $tRNA^{Phe}$ (Hingerty *et al.*, 1978) (heavy points) and, for comparison, the *deoxy*ribose puckers from the crystal structure of the self-complementary dodecamer $d(CGCGAATTCGCG)_2$ (Fratini *et al.*, 1982) (light points). (Copyright © 1986, American Chemical Society. Reprinted by permission.)

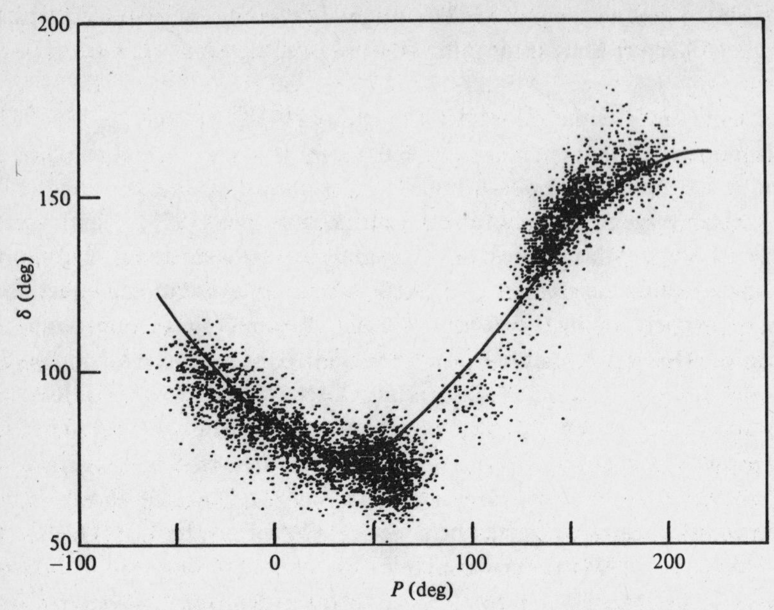

Fig. 6.7. Scatter plot showing the relationship between the exocyclic torsion δ (O3′-C3′-C4′-C5′) and the pucker phase angle P for 4800 ribose structures from a $tRNA^{Phe}$ molecular dynamics simulation (Harvey & Prabhakaran, 1986). The solid line represents the relationship of equation 6.2 plus the observation that, for ideal tetrahedral geometry at the C3′ and C4′ carbons, it would be expected that $\delta = \theta_1 + 120°$ (see figure 2.8). (Copyright © 1986, American Chemical Society. Reprinted by permission.)

cycle (Westhof & Sundaralingam, 1980; Olson, 1982; Olson & Sussman, 1982). The analysis of the 4800 dynamically generated ribose configurations from the tRNA simulation shows such variations, in addition to substantial thermal fluctuations in the bond angles.

The detailed local structural transitions accompanying the longitudinal stretching of DNA to form an intercalation site have been examined by both static and dynamic methods. Taylor & Olson (1983) generated all feasible combinations of the torsional angles of each nucleotide. They then followed the pathways that allow for smooth interconversion between structures that would allow intercalation. A dynamic approach was used by Prabhakaran & Harvey (1985), who forcibly stretched the double helix during a molecular dynamics simulation. This use of time dependent constraints takes advantage of the proven ability of molecular dynamics to anneal stereochemical conflicts (Brooks *et al.*, 1983; Levitt, 1983*c*).

The pathway for the B-DNA to Z-DNA transition has been examined in detail in two studies (Olson, Srinivasan, Marky & Balaji, 1983; Harvey,

1983). The vibrational entropy difference between the B and Z conformations has been calculated in a normal mode analysis (Irikura, Tidor, Brooks & Karplus, 1985); the calculated value, $91\ \mathrm{J\,mol^{-1}\,K^{-1}}$, is of the same order as the values determined by NMR (Cavailles *et al.*, 1984; Feigon *et al.*, 1984).

6.4 Nature of local structural transitions

Studies of fast motions in proteins and nucleic acids have shown that the atomic displacements resemble those observed in the liquid and solid states, as has been discussed in chapter 5. It is therefore useful to consider the general characteristics of local activated processes in liquids and solids as a reference point for the corresponding processes in biopolymers. Activated processes in condensed phases can be analyzed in terms of equations (4.29) and (4.30). Processes for which the transmission coefficient $\kappa = 1$ correspond to an ideal transition state theory limit. This theory assumes that the frictional effects (e.g., collision frequencies) within a system are large enough to insure that an equilibrium population of activated states is maintained in an ensemble of reacting systems, and that systems that have crossed the energy barrier are quenched in the final state region before they can rebound back across the barrier. It is also assumed that the frictional effects are small enough that systems crossing the barrier region in the direction of the final state are not interrupted and deflected back toward the initial state. These somewhat conflicting assumptions cannot be perfectly realized in any actual system, so that κ is always less than 1. Much work in chemical physics has been devoted to clarifying the nature of departures from the ideal transition state theory (Kramers, 1940; Chandler, 1978; Skinner & Wolynes, 1978; Northrup & Hynes, 1980; Grote & Hynes, 1980; Garrity & Skinner, 1983; Hynes, 1985).

In a liquid, the collision frequencies are sufficiently large that the rates of some processes may be reduced substantially by interruption of system trajectories in the barrier region. Such effects are most pronounced for the motion of large groups (which have large collision cross-sections and friction coefficients) over broad energy barriers. For a highly damped, one dimensional process in which the effective potential surface is parabolic in the initial, barrier and final state regions, the rate constant can be expressed as (Kramers, 1940; Chandrasekhar, 1943)

$$k = \kappa(\omega_i/2\pi)\exp(-\Delta W/k_\mathrm{B}T) \tag{6.3}$$

$$\kappa = \omega_\mathrm{b} m/f \tag{6.4}$$

Here, ω_i and ω_b are the angular frequencies of oscillation associated with the initial well and inverted barrier, respectively; ΔW is the height of the potential of mean force barrier on the reaction coordinate; k_B is Boltzmann's constant; T is absolute temperature; m is the mass of the particle moving along the reaction coordinate; f is the friction coefficient of the particle; and κ is a transmission coefficient which would be equal to one in the simple transition state theory. The high damping limit is obtained for certain conformational changes of flexible molecules in solution; the rate constant is reduced from the ideal transition state value because such molecules cross the energy barriers in an erratic, Brownian manner (Levy, Karplus & McCammon, 1979). The simple transition state picture may also break down for processes that involve motion of charged groups in polar solvents. Here, the motion of the charges may be strongly influenced by fluctuations in the polarization of the solvent (van der Zwan & Hynes, 1982; Warshel, 1982; Calef & Wolynes, 1983; Hynes, 1985), or by collective reorganization of hydrogen bonds in the solvation shells of the charged groups (Karim & McCammon, 1986). If one defines a primitive reaction coordinate in terms of the charge separation, the polarization or hydrogen bonding fluctuations will again lead to an erratic, Brownian motion along this coordinate. In extreme cases, the rate of the process will be controlled by the collective motions of the solvent molecules.

In solids, one again observes certain significant departures from simple transition state theory due to the nature of the underlying atomic motion. Consider, for example, vacancy diffusion in a simple monatomic crystal (Bennett, 1975; Becker & Hoheisel, 1982). Typically, an atom must squeeze between several nearest neighbors to move into the vacancy site. The necessary displacement of the neighbors tends to occur first and engenders strain in the lattice that provides the dominant contribution to the activation energy. The mobile atom then passes between its neighbors, experiencing much smaller nonbonded repulsions than would occur in the undistorted lattice. Detailed studies show that a significant number of barrier recrossings occur; the mobile atom retains enough kinetic energy to return to its initial site before the permissive configuration of the neighboring atoms collapses.

Activated process in proteins and nucleic acids can be expected to display many of the properties observed in liquids and solids as well as additional properties that result from the unique structural features of these molecules. Processes that involve small displacements of nonpolar groups are likely to be subject to relatively small frictional forces. The rotational isomerization of a methyl group, for example, can be expected to have a nearly ideal transition state character although some recrossing of the barrier may occur. Processes that involve large displacements of nonpolar groups will generally be more complicated in several respects. First, the group will be subject to direct frictional effects that act to reduce the rate constant. Second, the motion is likely to involve systematic coupling to collective distortions in the rest of the molecule. The covalent connectivity of the atoms in the molecule may result in more extensive collective distortions than would be seen, for example, in a corresponding small molecule crystal. This property is apparent in the tyrosine ring isomerization study described in section 6.2; in this process, the displacement of a sizable region of polypeptide backbone is coupled to the local rotation. The extensive character of the collective component leads to a significant time lag between the backbone and ring motions; the collective displacement gates the ring rotation. Experimental studies of aromatic ring rotations in dense, amorphous polymers suggest that these rotations are similarly gated by collective motion of the polymer chains (Schaefer *et al.*, 1985). Such effects complicate the interpretation of activation energies and other experimental parameters because the collective motion is likely to make an important contribution. To analyze such processes, it may be essential to consider their intrinsically multidimensional character (McCammon *et al.*, 1983; Aleksandrov & Goldanskii, 1984; Northrup & McCammon, 1984), as illustrated in figure 6.8. In proteins, relatively weak interactions between discrete structural domains may allow seemingly localized activated processes to be coupled to global motions of the molecule. For example, the transitions of groups located at the interface of two domains may be coupled to the relative motion of those domains. An extreme example is the coupling to ligand displacements in hinge bending proteins (cf. chapter 7). More subtle features of local–global coupling can result from the collective transitions that occur among the slightly different substates of a given native state (cf. chapter 5). The energy barrier for a local transition may be somewhat different in different substates, so that the local kinetics depend on the dynamics of these collective transitions (see below). In cases where the collective motion involves the protein surface, the rate of the local activated process is expected generally to display a dependence on solvent viscosity and other environmental factors that affect displacements of the surface (McCammon *et al.*, 1976; McCammon & Wolynes, 1977; Beece *et al.*, 1980; Northrup & McCammon, 1984).

Activated processes that involve displacements of charged groups can be expected to display all of the above features plus a dependence on polarization or hydrogen bonding fluctuations within the protein and the solvent. As with reactions that involve charge reorganization in polar

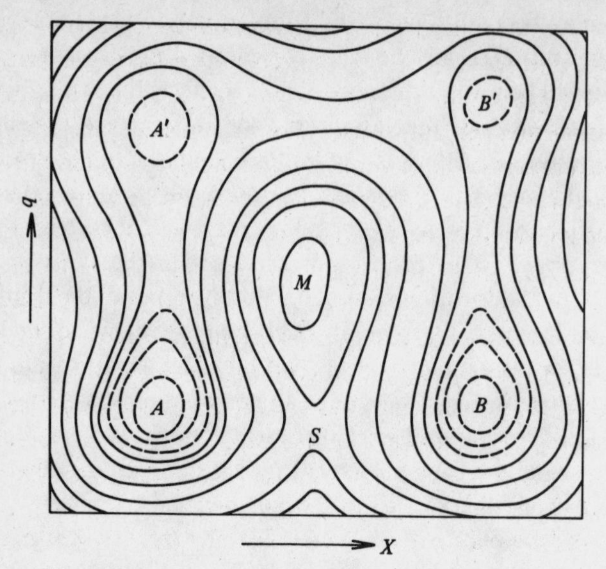

Fig. 6.8. Schematic representation of the potential surface corresponding to a simple gated reaction (Northrup & McCammon, 1984). The spatial coordinates used to describe the reaction are a primitive coordinate X and a gate coordinate q. The local minima A and B correspond to stable states within the gate-closed conformation; A' and B' are corresponding states within the gate-open conformation. M is a local maximum and S is a saddle point. Transitions from A to B will typically proceed through the intermediate states A' and B' rather than proceeding directly through the higher energy region around S. In many cases, motion along the primitive coordinate corresponds to local motion (e.g., ring rotation) while that along the gate coordinate corresponds to collective motion (e.g., displacement of a section of the protein adjacent to the ring). The overall rate of the A to B transition will therefore often reflect global influences (e.g. solvent viscosity, alterations of the protein rigidity that result from binding of effector molecules, etc.). (Copyright © 1984, American Chemical Society. Reprinted with permission.)

solvents, such reactions may be limited by the frequency of occurrence of permissive environmental polarizations; this represents another example of collective gating of local reactions.

As was stated at the beginning of this chapter, local structural transitions are important in the biological activity of proteins and nucleic acids. In this connection, it is of interest that the parts of biological molecules that are directly involved in such activity (e.g., ligand binding sites, catalytic sites in enzymes) often exhibit thermal displacements of particularly large magnitude (Artymiuk *et al.*, 1979; Frauenfelder *et al.*, 1979; Huber & Bennett, 1983). Local transitions of biological importance typically have activated character, with energy barriers due to steric hindrance, solvation changes, or electronic reorganization. The dynamic behavior of proteins, nucleic acids, and their solvent surroundings must be considered in any detailed discussion of these transitions. For example, the experimental rates of activated processes have generally been analyzed in terms of the thermodynamic formulation of transition state theory, in which the rate constant is expressed as (Fersht, 1985)

$$k = (k_B T/h) \exp(\Delta S^{\ddagger}/R) \exp(-\Delta H^{\ddagger}/RT) \tag{6.5}$$

Here, h is Planck's constant, R is the gas constant, $\Delta S^{\ddagger}$ is the entropy of activation and $\Delta H^{\ddagger}$ is the enthalpy of activation. The latter two quantities are determined by measuring the temperature dependence of the rate constant. Other similar quantities are determined by varying other conditions; e.g., measurement of the pressure dependence of the rate constant leads to an activation volume defined by

$$\Delta V^{\ddagger} = -k_B T \left(\frac{\partial \ln k}{\partial \ln p} \right)_T \tag{6.6}$$

Although this approach allows a useful reduction of experimental data in terms of a small number of parameters, recent studies have shown that such parameters must be interpreted with caution in any attempt to make contact with the microscopic dynamics of the system. Thus, $\Delta S^{\ddagger}$ and $\Delta H^{\ddagger}$ will not be simply related to the entropy and enthalpy costs of crossing a barrier if the actual enthalpy barrier exhibits a significant temperature dependence. Such a temperature dependence can arise, for example, from thermal expansion and contraction effects that modulate steric hindrance in dense materials (Northrup *et al.*, 1982*a*). Similarly, an activation volume does not in general correspond to a physical volume that must open up to allow an activated process to proceed (Montgomery, Chandler & Berne, 1979; Karplus & McCammon, 1981*b*). Because of the intimate coupling between local and collective motions in biopolymers, caution is also required in attempts to interpret parameters such as $\Delta H^{\ddagger}$ (or even the actual enthalpy barriers) in local terms (Northrup & McCammon, 1984). For example, collective polarization fluctuation may play a role in certain of the catalytic steps of lysozyme (Warshel & Levitt, 1976; Warshel, 1984).

A second type of difficulty in the simple thermodynamic transition state theory approach arises from the neglect of nonequilibrium effects in (6.5). Frictional effects generally decrease the rate of a reaction from what would be expected based on the potential energy surface alone and, in some cases, may even determine the dominant mechanism or pathway of reaction (cf. chapter 7). That such effects are important in biology is clear, for example, from experimental studies of ligand binding to myoglobin (Beece *et al.*, 1980). As described in section 6.2, analysis of such experiments suggests that the ligand can occupy different stable locations within the protein before exiting to the solvent or binding to the iron. The stable locations are separated by energy barriers that are partly of steric origin and that fluctuate in magnitude as a consequence of the normal internal motion of the protein. The experiments show that the rate of hopping over each barrier depends on the solvent viscosity, even for barriers corresponding to transitions within the protein interior. These results are most simply understood in terms of a coupling of the local ligand displacements to collective fluctuations that involve motion of the protein–solvent interface. In the language used elsewhere in this book, the ligand hopping can be viewed as a gated process and the dynamics of the gate are influenced by the viscosity of the solvent.

7

Global structural changes

7.1 Introduction

The biological importance of large scale intramolecular motions is obvious. Such motions are an essential part of protein folding, the binding of ligands by proteins, and allosteric effects in enzymes. They are important in a variety of interactions between macromolecules, including the aggregation of antibodies (Yguerabide, Epstein & Stryer, 1970; Hanson, Yguerabide & Schumaker, 1985), the formation of the protein coat of viruses (Harrison, 1978), muscle contraction (Huxley, 1969; Harrington, 1971; Harvey & Cheung, 1982; Eisenberg & Hill, 1985), and the packaging of DNA in the nucleosome (Olins & Olins, 1974; Kornberg, 1974; Levitt, 1978; Sussman & Trifonov, 1978). They may be coupled to local intramolecular motions of the kind described in chapter 6, for example in the repuckering of sugars accompanying the transitions between the A, B and Z conformations in DNA, or in the formation of the environments necessary for some local structural transitions.

Motions that involve most or all of the atoms in a macromolecule occur over a wide range of time scales. The fastest of these motions are small amplitude vibrations whose characteristic times in the absence of solvent (determined from normal mode calculations) may range up to 10 ps. When the effects of solvent are included, the time scale can increase substantially, because the motions are no longer free, undamped oscillations. The time scale will depend on such factors as the masses of the molecular fragments, the elastic force opposing the motion, the friction due to the solvent, and the relative magnitudes of the inertial and viscous forces in the solvent dynamics (the Reynolds number). Even in cases of relatively small solvent effects, the time scale will more than double (McCammon *et al.*, 1976; McCammon & Wolynes, 1977). Very large scale segmental motions that are subject to weak restoring forces may be so highly overdamped that they are purely diffusive, being described by classical theories of Brownian motion (Chandrasekhar, 1943); such motions typically have time scales of nanoseconds or longer. The local denaturation of chain segments may involve the diffusional exploration of a large number of low energy conformations; the lifetimes of such denatured states may range from tens of nanoseconds to more than a microsecond (Englander & Kallenbach, 1984; Creighton, 1985).

The type of method used to study these slow, extensive motions depends on the complexity of the motion of interest. For small amplitude, low frequency vibrations, normal mode analyses (section 4.6) can identify some of the important features. Although quantitative errors can arise from anharmonic effects, such errors do not appear to be very large in the case of small amplitude motions (cf. section 5.2.2). Other methods for studying such motions include the quasiharmonic approaches, in which one considers slow oscillations near the minima of a potential surface that has been averaged with respect to the fast motions of a system. There are several ways to carry out this averaging. One can take the familiar approach of continuum mechanics, in which the system is characterised by suitable elastic constants and damping parameters. The elastic constants describe the strain in the material in response to a slowly varying applied stress and therefore implicitly include the averaging with respect to fast motions. This approach has been used to characterize some of the low frequency oscillations of a spherical protein (de Gennes & Papoular, 1969; Suezaki & Gō, 1975). A second approach is to monitor the large scale displacements that occur in a full molecular dynamics simulation, and to use this information to construct a potential of mean force for appropriate collective coordinates in the molecule (Levy *et al.*, 1984*c*). The resulting temperature dependent potential typically has a quadratic minimum in the region of the native structure and can be used as a basis for effective normal mode calculations.

In cases where it is possible to focus on a specific mode or displacement path, but where the motions may be of large amplitude, there are three potentially useful methods. The first is the adiabatic mapping technique, in which one displaces the system along the path of interest while minimizing the energy of the other degrees of freedom (section 4.5). The resulting potential may provide an approximation to the potential of mean force for displacements along the path. Given the effective potential surface, the dynamics of the motion in the presence of solvent surroundings can then be modeled by use of the Langevin equation (McCammon *et al.*, 1976; McCammon & Wolynes, 1977). A second approach is to deform the system along the path of interest during a molecular dynamics simulation, using the simulated thermal motions to anneal the stereochemical stresses that build up during the deformation. Although this does not produce a true representation of the time course of the global motion, it can reduce the energetic cost of the deformation to a fraction of that which would be determined by the adiabatic method (Prabhakaran & Harvey, 1985). A third approach, which has not been applied to date, would be to calculate the potential of mean force more rigorously using the umbrella sampling method (section 4.8). To do this would require the selection of a set of representative initial structures at regular intervals along the deformation pathway; these could be generated by the adiabatic mapping method, by the method of forced deformation with annealing by molecular dynamics, or by choosing the initial configuration for a given umbrella simulation from a previous simulation that explored an adjacent section of the deformation pathway.

For large amplitude motions where one is not able to specify the path in advance, it may be useful to apply Brownian dynamics simulation methods (section 4.10). This is the diffusional analogue of molecular dynamics, carried out by the numerical integration of the Langevin equation. The potential function is simplified by averaging over the rapid motions in the molecule and over displacements of solvent molecules and other groups whose detailed motions are not of interest. The degrees of freedom that are eliminated by this averaging process are included in a thermal bath that contains both frictional and random forces. For extensive, activated processes where the motion involves barrier crossing, special methods analogous to those used for inertial systems are useful (Northrup & McCammon, 1980*b*).

7.2 Results for proteins

Protein molecules exhibit a variety of large scale motions. Motions that can be described as collective fluctuations within a single globular domain have already been discussed in section 5.2.2. For the small protein BPTI, such motions involve oscillations with frequencies as low as 0.1 ps^{-1} and transitions among similar structures that have lifetimes as long as 40 ps and probably much longer. Individual atoms may be displaced by 0.2 or 0.3 nm in such motions, but the relative displacements of nearby atoms are usually significantly smaller. In the remainder of this section, we will consider two other types of large scale motion, namely hinge bending motions and single strand motions. These types of motion can produce rather large relative displacements of atoms within a protein and are involved in the biological activity of a large number of proteins.

Many enzymes and ligand binding proteins have two or more globular domains connected by relatively flexible strands of polypeptide (Ptitsyn, 1978; Richardson, 1981; Janin & Wodak, 1983; Bennett & Huber, 1984; Johnson, 1984; Petsko & Ringe, 1984). The binding sites for substrate or ligand are typically located in a cleft between two such domains, so that the relative motion of the domains is likely to play an important role in binding and subsequent activity. Similar hinge bending motions are important in the function of antibodies and in myosin, a muscle protein.

The first protein for which a dynamical analysis of hinge bending motions was attempted is lysozyme, a small enzyme with two globular domains separated by an active site cleft (McCammon *et al.*, 1976). An approximate potential for the hinge bending was calculated by an adiabatic mapping procedure. A bending axis was first identified by examination of the crystal structure of the protein. One lobe of the protein was then rigidly rotated through various angles about this axis. After each rotation, the outer parts of the two lobes were held fixed and the remainder of the protein was relaxed by use of an energy minimization procedure. This relaxation process relieved local strains (e.g., close contacts between nonbonded atoms and bond angle deformations). Because the characteristic time scale for these local motions (less than 0.1 ps) is much shorter than the time scale for the hinge bending motion itself (greater than 10 ps, see below), the resulting adiabatic potential should be a reasonable approximation to the effective potential for the hinge bending. The effective potential is approximately quadratic in shape. An analysis of the calculated stress–strain ratio in terms of the equations for the deformation of a thick, linear beam shows that the Young's moduli are of the order of 10^{-11} dyn cm^{-2}, in the range expected for proteins (Karplus & McCammon, 1981*a*), and that bending dominates shear by about an order of magnitude in determining the stiffness of the protein. The calculated force constant is such that typical fluctuations of the hinge angle of the order of 3° are expected at room temperature; the corresponding changes in width of the opening of the cleft are of the order of 0.1 nm. Quite recently, nearly identical force constants have been obtained by means of normal mode calculations (Brooks & Karplus, 1985) and by improved adiabatic mapping calculations (Bruccoleri, Karplus & McCammon, 1986). The latter calculations also indicate that, upon binding of the inhibitor tri-N-acetyl-glucosamine, the hinge closes by 10° and the force constant increases by a factor of two.

Because the hinge bending motion of lysozyme involves a substantial movement of the protein surface, it is essential to consider the solvent in a description of the hinge bending dynamics (McCammon *et al.*, 1976; McCammon & Wolynes, 1977). This was done within the framework of the harmonic oscillator Langevin equation by calculating the effective

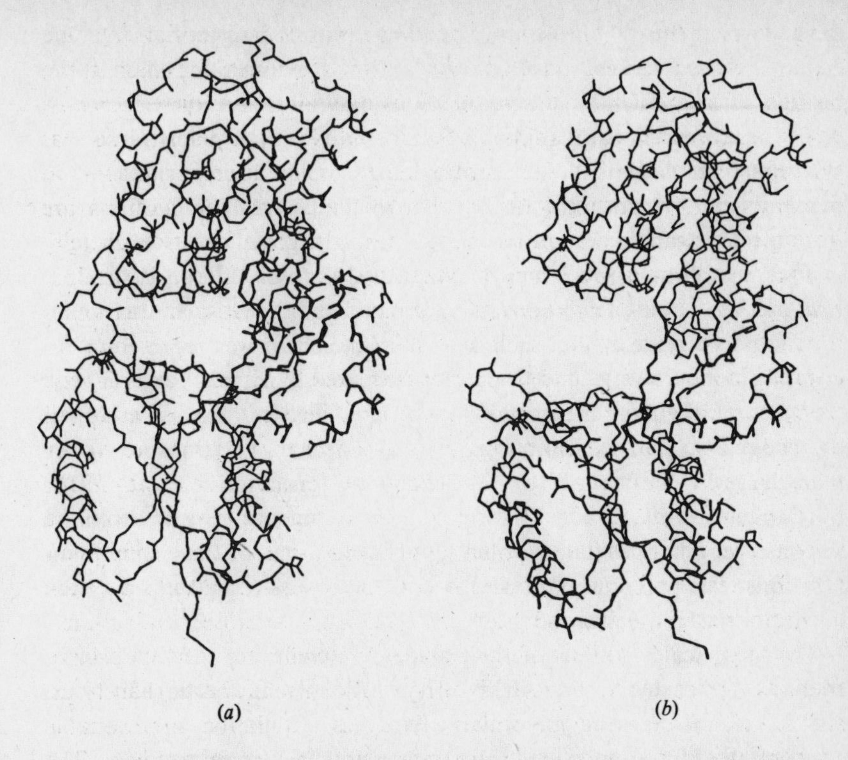

Fig. 7.1. Structures of the L-arabinose-binding protein corresponding to hinge bending angles of 0° (*a*) and 28° (*b*) (Mao *et al.*, 1982*b*). Sidechain atoms are omitted.

moment of inertia for the protein bending and the friction constant associated with solvent damping. The damping was treated by modeling the two globular domains as spheres and calculating the viscous frictional drag opposing the relative motion of these spheres by use of a modified Stokes law. The relative motion of the two globular domains in lysozyme was found to be overdamped, with a characteristic relaxation time of 20 ps. Thus, the hinge bending motion is expected to be Brownian in character. A typical fluctuation will open or close the cleft by about 0.1 nm and will persist for approximately 20 ps.

Much larger hinge bending motions occur in other proteins. For the L-arabinose-binding protein, calculations similar to those described above indicate that the binding cleft, which is open in the unliganded state, is induced to close upon ligand binding; the two lobes of the protein swing through a relative angle of approximately 30° upon closing (Mao *et al.*, 1982*b*; Mao & McCammon, 1983, 1984). The open and closed structures of the protein are illustrated in figure 7.1.

Certain large scale motions in proteins involve structural changes that are less well defined than the hinge bending motions described above. One example is the transient unfolding of a region of polypeptide chain at the protein surface. Such local denaturations may allow complete rotations of the backbone dihedral angles in the displaced polypeptide strand. Approximate simulations of such motions are possible using the Brownian dynamics method (section 4.10). The Brownian dynamics analysis of large scale motions can be illustrated by studies of the growth of alpha helices in aqueous solution (McCammon, Northrup, Karplus & Levy, 1980; Pear *et al.*, 1981). These calculations made use of a simplified structural model for the polypeptide chain. Each residue in the chain was represented by a single interaction site, and the sites were linked by virtual bonds (Flory, 1969). A set of energy parameters for the simplified model was developed by recognizing that local motions such as sidechain rotations are faster than large scale changes in the chain conformation (Levitt, 1976; McCammon *et al.*, 1980). This separation of time scales was exploited to construct an approximate potential of mean force for the virtual bond rotations. The potential of mean force includes the average effects of the local motions and solvent interactions.

The large scale motions of the polypeptide chain are subject to large frictional forces due to the high viscosity of the solvent. Inertial effects are negligible, so that the Brownian dynamics simulation approach is appropriate (McCammon *et al.*, 1980). To study the dynamics of helix–coil transitions in polyvaline, a series of simulations was produced, each starting at the all-helix configuration. The diffusional motion of residues out of and back into the helical configuration was monitored. A pronounced end effect was observed; the rate constants for helix–coil transition of the terminal residue were about two orders of magnitude larger than those for residues closer to the middle of the chain. A more detailed analysis of the simulations revealed that adjacent residues often move from the helix to the coil or *vice versa* in nearly concerted fashion (Pear *et al.*, 1981). Such transitions are not consistent with the conventional idea that successive transitions occur independently (Schwarz & Engel, 1972). The frequent occurrence of the correlated transitions results from the relatively small frictional forces associated with these particular motions, which avoid large displacements of the end of the chain through the solvent (Pear, Northrup & McCammon, 1980; Pear *et al.*, 1981). The trajectories of two such transitions are shown in figure 7.2.

Large scale internal motions of protein molecules have been studied by a variety of experimental methods. In crystals of proteins that have globular lobes connected by hinges, the pattern of crystallization may be such as to produce mechanical anisotropy. This has been demonstrated

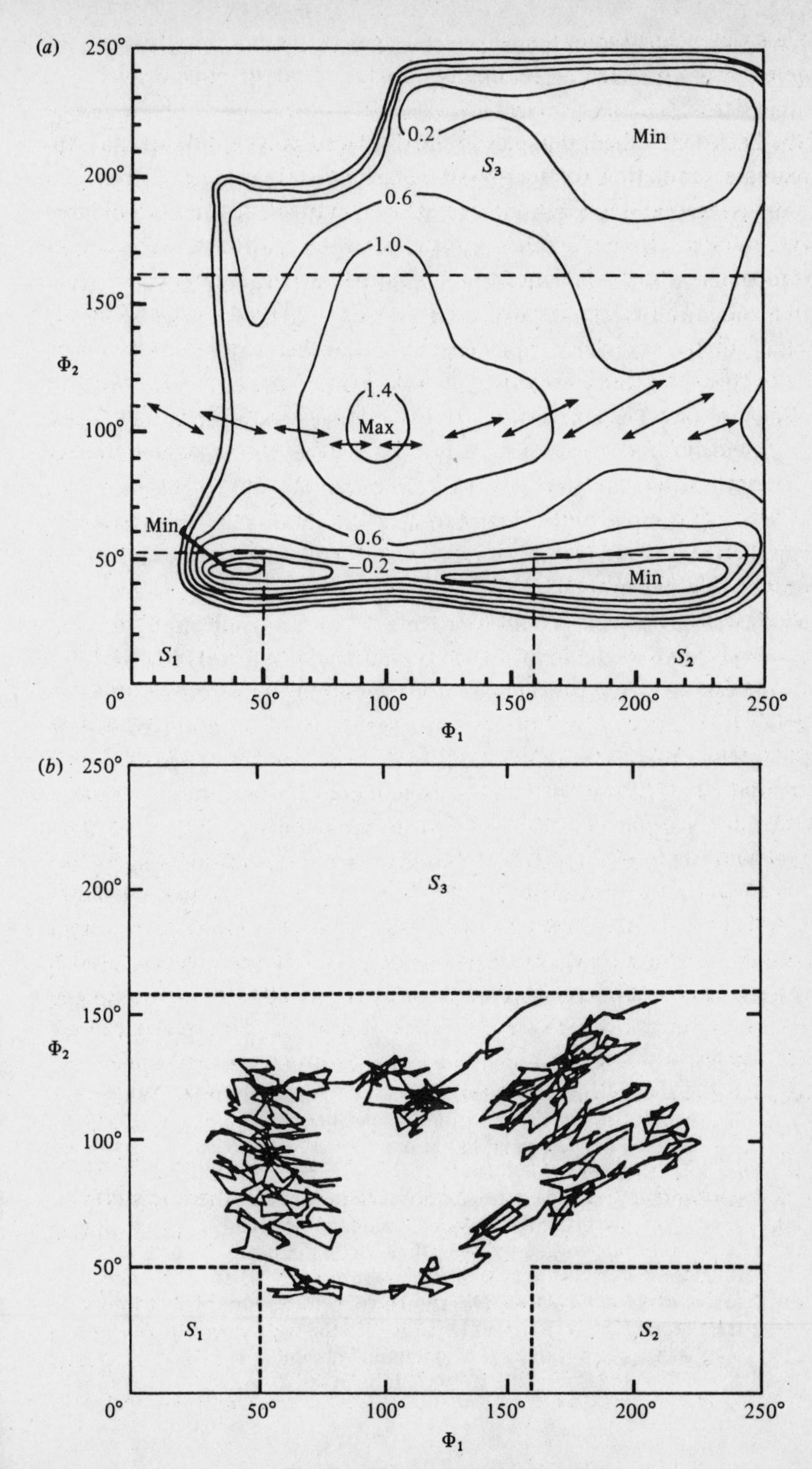

Fig. 7.2. (for caption see overleaf).

for triclinic lysozyme crystals, which are relatively easily deformed by compressive forces that act in the direction corresponding to hinge bending of the molecules in the crystal (Morozova & Morozov, 1982). The apparent flexibility of the individual molecules is roughly consistent with the theoretical prediction (McCammon *et al.*, 1976). Hinge bending flexibility is also apparent in comparisons of high resolution X-ray structures of certain proteins (Janin & Wodak, 1983; Bennett & Huber, 1984; Johnson, 1984; Petsko & Ringe, 1984). Although such comparisons generally do not provide quantitative mechanical or dynamical measures of the molecular flexibility, they do provide qualitative indications of the types of deformations involved and important suggestions as to the biological function of such motions. The first demonstration of hinge bending flexibility was in fact provided by a comparison of lysozyme structures with and without an inhibitor bound in the active site (Imoto *et al.*, 1972); the two lobes of the enzyme move toward each other through a relative distance of about 0.1 nm when the inhibitor is bound. Much larger motions have been observed in other proteins.

Studies of proteins in solution provide data on hinge bending that are not complicated by the influence of crystal packing forces, but here one loses the detailed structural information provided by high resolution X-ray analysis. Ligand-induced decreases in the radii of gyration of several proteins have been observed in small angle X-ray scattering studies (Janin & Wodak, 1983). For the **L**-arabinose-binding protein, the decrease of about 0.1 nm upon binding of **L**-arabinose is consistent with the large hinge bending motion described above and illustrated in figure 7.1 (Newcomer, Lewis & Quiocho, 1981). The time scales of hinge bending oscillations (but not ordinarily of ligand-induced transitions) can be determined in fluorescence depolarization or inelastic neutron scattering experiments. The former technique reveals large motions of the domains

---

Fig. 7.2. Configuration space defined by the virtual dihedral angles for the two residues at the end of an α-helical region of polyvaline (Pear *et al.*, 1981). (*a*) The potential of mean force with contour intervals of 0.4 kcal/mol (1 kcal = 4.184 kJ). The minima at S_1, S_2, and S_3 correspond respectively to the most stable configurations with all residues in the helix, the endmost residue out of the helix, and the two residues at one end out of the helix. Arrows indicate the preferred directions of diffusion (displacements subject to the smallest frictional forces in the solvent). (*b*) Diffusional trajectories for two helix–coil transitions. The trajectories begin just to the right of region S_1. Neither trajectory follows the commonly assumed sequence $S_1 \rightarrow S_2 \rightarrow S_3$. (Copyright © 1981, John Wiley & Sons, Inc. Reprinted by permission.)

of immunoglobulins, with reorientation times in the range of 10–50 ns (Yguerabide *et al.*, 1970; Holowka & Cathou, 1976); these results are in accord with theoretical estimates based on a diffusional hinge bending model for the domain motions (McCammon & Karplus, 1977). Fluorescence depolarization studies detect similar motions of the myosin head (Mendelson, Morales & Botts, 1973). By contrast, the putative hinge in the myosin rod is found to be quite stiff when examined by this technique (Harvey & Cheung, 1977, 1980, 1982). Results from inelastic neutron scattering, although less directly interpretable than those from fluorescence depolarization, suggest the presence of hinge bending motions in the enzyme hexokinase and the suppression of these motions upon substrate binding (Jacrot, Cusack, Dianoux & Engelman, 1982).

Turning to single strand motions, a variety of relevant experimental data is again available. Evidence from high resolution X-ray diffraction studies has confirmed the occurrence and biological importance of strand folding–unfolding transitions in several proteins (Bennett & Huber, 1984; Petsko & Ringe, 1984). Such transitions typically involve protein segments near ligand binding or active sites. Upon ligand binding, these segments assume ordered conformations in which they cover or otherwise interact with the bound ligand. An interesting variation on this theme is provided by the activation domain of the enzyme trypsin and its inactive precursor, trypsinogen. The activation domain comprises several loops of polypeptide that are disordered in the latter molecule but are induced (by interaction with the amino terminus) to assume a well-defined structure in the active enzyme. The dynamics of the activation loops of trypsinogen have been studied by analysis of the perturbed angular correlation of mercury atom labels (Butz, Lerf & Huber, 1982). The apparent characteristic time for the loop motion is about 11 ns, which is comparable to the time required for diffusional reorganization of polypeptide chains in Brownian dynamics simulation studies (McCammon *et al.*, 1980). The latter studies can also be compared to experimental data on the growth kinetics of alpha helices where, again, reasonable accord is found (McCammon *et al.*, 1980).

7.3 Results for nucleic acids

7.3.1 *Large scale motions in DNA*

One of the earliest computer modeling studies of nucleic acid structure using an atomic model was that by Levitt (1978), in which energy minimization was used to optimize models of a 20 basepair segment of DNA. In the models, Levitt varied the global helix twist angle, asking how many basepairs per helical turn would produce the lowest energy structure. The calculations were carried out both for a linear DNA geometry and for the case where the axis of the double helix was bent smoothly into a radius of 4.5 nm, appropriate to the diameter of the nucleosome core particle (Kornberg, 1974; Olins & Olins, 1974). It was found that, for a fragment of random sequence but with equal numbers of A-T and G-C basepairs, the predicted optimum helix twist in the curved segment would be about ten basepairs per turn, compared with a predicted optimum of about 10.5 basepairs per turn in the straight segment (Levitt, 1978). This result was invoked as an explanation of the differences in helix repeat for DNA free in solution and DNA in nucleosomes; the differences had been detected experimentally in nuclease digestion patterns (Noll, 1974) and in changes in linkage number (Germond *et al.*, 1975; Keller, 1975). This study was also significant because, in addition to rationalizing the results of previous experiments, it predicted that basepairs would deviate from the planar configuration of the classical models that had been derived from fiber X-ray diffraction studies (Watson & Crick, 1953; Arnott *et al.*, 1976). Levitt (1978) predicted pronounced basepair tilting, nonzero roll angles, and propeller twisting, all of which were subsequently confirmed by single crystal studies (Wing *et al.*, 1980; Fratini *et al.*, 1982).

The first molecular dynamics simulation of DNA, which treated the self-complementary dodecamer d(CGCGAATTCGCG), produced spontaneous periodic bending oscillations, with structures that varied from a nearly straight segment to one with a radius of curvature of 2.0 nm (Levitt, 1983*d*). The period of the oscillation, 26 ps, is similar to that of the lowest frequency bending mode of a DNA model of similar size, determined by a normal mode analysis (Tidor *et al.*, 1983). Both the molecular dynamics calculation of Levitt (1983*d*) and the normal mode analysis of Tidor *et al.* (1983) were done for models *in vacuo*, and it would be expected that this bending mode would be very overdamped in solution.

The full treatment of overdamped bending modes for long double helical pieces of DNA is a suitable problem for simulation by the method of Brownian dynamics (section 4.10). Such a simulation has recently been carried out (Allison & McCammon, 1984*b*). The DNA chain is modeled as a string of beads, each bead representing a section of the double helix 3.2 nm in length. The calculation includes terms to account for the full effects of hydrodynamic interactions and to reproduce the bending stiffness of DNA, with a persistence length of 60 nm. The simulations amount to computer experiments, and because a time step of 10 ps can be used, it was possible to follow the dynamic evolution of an ensemble of 100 randomly generated chain configurations for periods ranging from 3 to 200 ns (Allison & McCammon, 1984*b*). The data were used to generate model fluorescence polarization decay curves and model decay curves for the structure factor from polarized dynamic light scattering. The former compared very well to the fluorescence depolarization predicted by the theory of Barkley & Zimm (1979), while the light scattering results were found to be insensitive to the internal motions of DNA in the low scattering angle limit. These results emphasize how different experiments probe different dynamic as well as conformational aspects of the same system (Allison & McCammon, 1984*b*).

7.3.2 *Large scale motions in tRNA*

As soon as the first crystal structure of $tRNA^{Phe}$ became available, it was suggested that the L-shaped molecule might be hinged between the D stem and the anticodon stem (Robertus *et al.*, 1974). Based on a nonmonotonic dependence of the translational diffusion coefficient on ionic strength, a second possible hinge was proposed between the two arms of the molecule, in the region of P8 (the phosphorus atom of the eighth residue) and P49 (Olson, Fournier, Langley & Ford, 1976). Several other investigations have examined the question of hinge bending in $tRNA^{Phe}$ (Reid, 1977; Schwarz, Möller & Gassen, 1978; Wickstrom, Behlen, Reuben & Ainpour, 1981; Nilsson, Rigler & Laggner, 1982), and a recent review has considered the evidence for a variety of motions in tRNA (Rigler & Wintermeyer, 1983). Indirect evidence for hinge bending comes from the crystallographic structure of $tRNA^{Asp}$ (Westhof, Dumas & Moras, 1985), where it was found that the two arms form an angle about 15° larger than that observed in $tRNA^{Phe}$. The putative hinge bending motion of $tRNA^{Phe}$ has been investigated by the adiabatic mapping method (section 4.5). In the initial study (Harvey & McCammon, 1981), the molecule was divided into two subfragments, and a series of small, rigid body rotations, followed by refinement with energy minimization after each rotational step, was used to bend the molecule about the proposed hinge at P8–P49. A total bending angle of 36° was achieved for motions that would open or close the 'L', and for twisting motions that would swing the upper arm forward and backward while keeping the two arms more or less perpendicular to one another. The deformation energies for these motions were in the range 0–200 kJ/mol; certain bending and twisting displacements up to about 10° were possible with deformation energies less than 10 kJ/mol. Refinements of the method (Harvey & McCammon, 1982; Tung *et al.*, 1984) led to a more extensive study that examined bending about other possible hinges and that included distributed flexibility (deformations that bend a double helical segment along a smooth curve). According to those calculations (Tung *et al.*, 1984), $tRNA^{Phe}$ can be opened up into a nearly linear conformation where the

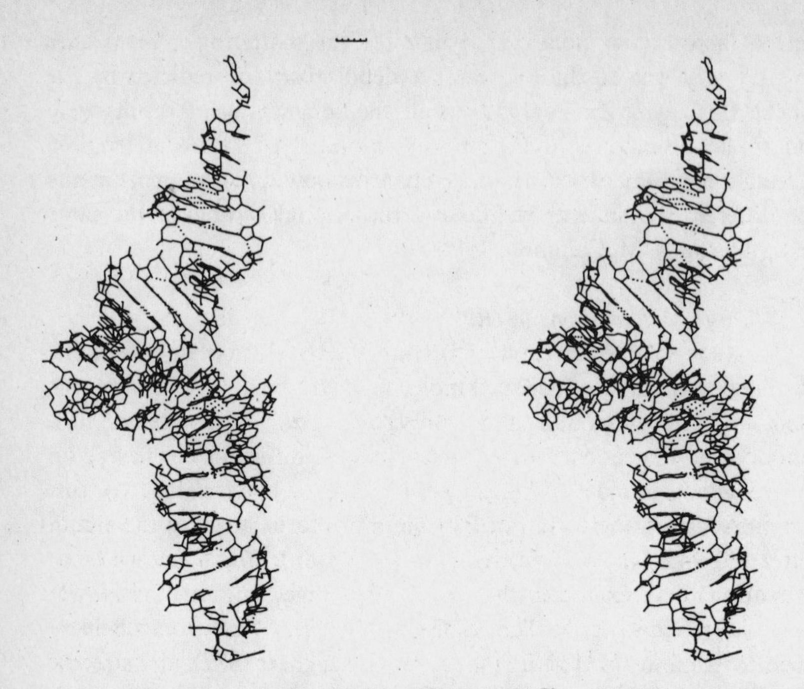

Fig. 7.3. Fully extended model of $tRNA^{Phe}$, obtained by bending the crystallographic model of the molecule, using the method of adiabatic mapping (Tung *et al.*, 1984). Scale bar is 0.5 nm. (Copyright © 1984, John Wiley & Sons, Inc. Reprinted by permission.)

separation between the anticodon and the 3′-acceptor terminus is about 9.0 nm; the energetic cost is about 200 kJ/mol (figure 7.3). When the molecule is doubled over so that the anticodon and the amino acid acceptor terminus are brought together, full closing of the molecule was not achieved, but the separation between the ends of the molecule could be reduced to about 4.0 nm at a cost of about 400 kJ/mol (figure 7.4). These energies are believed to be upper bounds on the actual deformation energies, because of the inefficiencies of energy minimization methods, as discussed in sections 4.4 and 4.5. It therefore appears likely that variations in solvent conditions and interactions with other molecules may produce large changes in the overall conformation of tRNA (Tung *et al.*, 1984).

Experimental information on the time scale of these hinge bending motions in $tRNA^{Phe}$ in solution is not yet available. The *in vacuo* molecular dynamics simulation (cf. section 5.3) did show small amplitude variations in the radius of gyration that arise from fluctuations in the bend angle between the two arms (Harvey *et al.*, 1985*a*). The length of the simulation (24 ps) is comparable to the apparent period of those motions, and it would

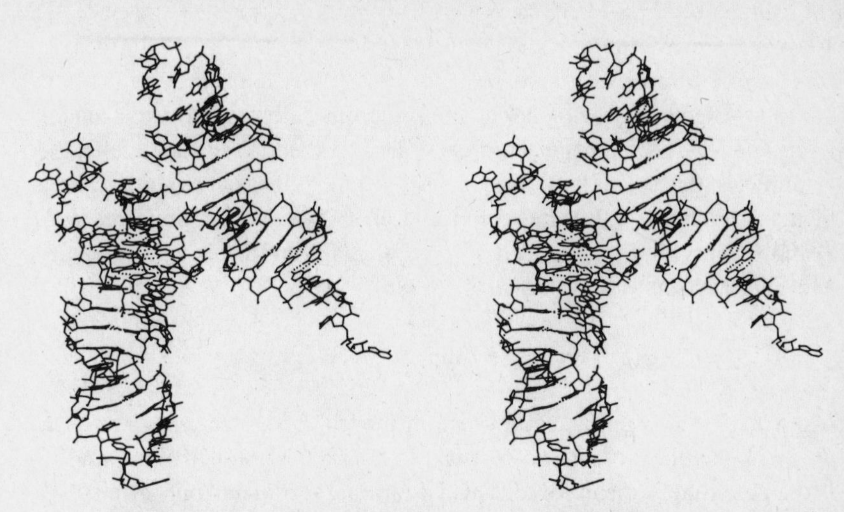

Fig. 7.4. Closed model of $tRNA^{Phe}$ from the adiabatic mapping study (Tung *et al.*, 1984). Scale bar is 0.5 nm. (Copyright © 1984, John Wiley & Sons, Inc. Reprinted by permission.)

be expected that solvent damping effects would be large enough to give characteristic times that approach the nanosecond range.

7.4 Nature of global structural changes

Because large scale motions typically involve substantial displacements of the protein or nucleic acid surface, the coupling between the biopolymer and its solvent surroundings is an important determinant of the nature of such motions. This coupling will, in general, influence both the probabilities of occurrence of different configurations of the biopolymer and the rates of transition among these configurations.

The simplest kinetic effect of the solvent is the slowing of internal motions of the biopolymer due to viscous damping. For lysozyme, the lifetime of an open conformation of the active site cleft is increased from about 8 ps for the underdamped motion expected *in vacuo* to about 20 ps for the overdamped motion in water at room temperature (McCammon *et al.*, 1976). The hinge bending motions in antibody molecules involve smaller restoring forces and larger frictional forces; the characteristic times for these motions are on the order of 10 ns (Yguerabide *et al.*, 1970; McCammon & Karplus, 1977). For DNA double helices, the time scale of undamped bending for segments 4–7 nm long would be on the order of 20–30 ps (Levitt, 1983*d*; Tidor *et al.*, 1983). The time scale of solvent damped bending increases with the length of the bent portion of the helix; substantial bending of 30 nm segments occurs in times on the order of 10 ns (Barkley & Zimm, 1979; Allison & McCammon, 1984*b*). In the case of motions that have an intrinsically multidimensional nature, solvent damping effects may also change the preferred pathways of displacement from what would be expected based on the potential surface alone. An example is the solvent induced correlation in dihedral angle rotations displayed in figure 7.2. It has recently been shown that the preferred path P of displacement for a highly damped system is that for which (Berkowitz, Morgan, McCammon & Northrup, 1983)

$$\int_P f_{tt} \exp(\beta W) \mathrm{d}s = \text{minimum} \tag{7.1}$$

where f_{tt} is the tangential component of the friction tensor along the path, $\beta^{-1} = k_B T$ is the Boltzmann constant multiplied by temperature, and W is the potential of mean force. This variational formula explicitly displays the dependence of the pathway of conformational change upon both frictional and potential effects. As will be discussed in the next chapter, such effects may influence the mechanistic details of ligand binding and other biological functions of biopolymers.

In the analyses of damping effects described above, the solvent has been treated as a viscous continuum. This simple model must be replaced by more sophisticated descriptions of the solvent in certain cases. For example, the large restoring forces involved in the lysozyme hinge bending result in sufficiently rapid motions that the simple viscous response model for the solvent begins to break down (McCammon & Wolynes, 1977). The solvent can still be modeled as a continuum, but one which has an inertial as well as viscous character. Other extensions of the continuum model are possible in principle, e.g., incorporating solvent dielectric features as has been done in studies of electrolyte solutions (Wolynes, 1980). In some cases, it may be necessary to introduce a detailed molecular model for the solvent. This may be required for biopolymer motions that result in significant alterations of the surface exposure. Such motions will involve changes in the solvation of the surface groups with concomitant energetic and kinetic effects. An approximate consideration of such effects upon the hinge bending motions of the L-arabinose-binding protein suggests that solvation changes of charged sidechains in the binding site cleft may regulate the overall conformation of this protein (Mao *et al.*, 1982*b*).

8

Dynamics of molecular associations

8.1 Introduction

As discussed in section 3.4, the association of enzyme and substrate molecules is influenced by a variety of inter- and intramolecular processes. These processes, which may include conformational changes in one or both species in addition to the diffusional encounter and desolvation steps, are involved also in the association of antigens and antibodies, hormones and receptors, and other molecular pairs that have a ligand–receptor relationship.

In most biochemistry textbooks, the rate of diffusional encounter between ligands and receptors is discussed within the framework of the traditional Smoluchowski theory. In this theory, the reaction partners are assumed to be spherical, with uniformly reactive surfaces. Apart from their instantaneous reactivity upon contact, interactions between the reaction partners are ignored. For the enzyme–substrate association

$$\mathrm{E+S} \xrightarrow{k} \mathrm{ES} \tag{8.1}$$

and typical molecular sizes, the Smoluchowski theory yields a bimolecular rate constant k of about $10^{10}\ \mathrm{M^{-1}\,s^{-1}}$ (Cantor & Schimmel, 1980; Fersht, 1985). It is frequently supposed that bimolecular reactions with rate constants more than about an order of magnitude below this value are not diffusion controlled. Of course, bimolecular encounters are in fact much more complicated than what is pictured in the Smoluchowski theory. Typically, only a small part of the molecular surface is active. Electrostatic and other interactions between the reaction partners will generally produce attractive or repulsive forces that can increase or decrease the rate of diffusional encounter. Structural fluctuations in the ligand or receptor may result in a time dependent reactivity upon contact. Effects such as these influence biological processes in a variety of interesting ways. For example, ligands and receptors that have complementary distributions of electric charge may be steered into productive orientations during diffusional encounters, resulting in increased reaction rates. In such cases, it can be said that molecular recognition begins before the ligand and receptor actually come into contact.

Because of the complications outlined above, reactions between biological molecules may be diffusion controlled and yet have rate constants that differ by several orders of magnitude from what is expected on the basis of the simple Smoluchowski theory. Analytic theories that take some of these complications into account have been developed and are described in several recent reviews (Chou & Zhou, 1982; Shoup & Szabo, 1982; Calef & Deutch, 1983; Berg & von Hippel, 1985; Keizer, 1985). However, it appears likely that the most detailed descriptions of diffusional bimolecular encounters and subsequent steps in molecular association will be provided by computer simulation methods of the kind described in chapter 4. An excellent review of the background and prospects for such studies has been provided by Dickinson (1985).

8.2 Superoxide dismutase

The diffusion controlled reaction of superoxide (O_2^-) catalyzed by the enzyme superoxide dismutase (SOD) is anomalous in at least two respects (Cudd & Fridovich, 1982; Getzoff *et al.*, 1983). First, the reaction rate ($k \approx 2 \times 10^9\ M^{-1}\ s^{-1}$ at $T = 25$ °C and ionic strength of 0.02 M) is larger than what one might expect, given that (a) the reactants have net electrostatic charges of the same sign (-4 for SOD and -1 for O_2^-, in units of proton charge) and hence suffer some electrostatic repulsion, and (b) the two specific binding pockets for substrate account for only 0.1% of the surface area of the dimeric enzyme. Second, the reaction rate decreases with increasing ionic strength of the solvent at moderate salt concentrations, whereas the opposite behavior would be expected for electrostatically repelling reactants. Chemical and structural studies of SOD suggest that these anomalies arise from an asymmetric charge distribution on SOD that acts to steer O_2^- into the active sites (Cudd & Fridovich, 1982; Getzoff *et al.*, 1983).

Allison and co-workers have initiated a series of Brownian dynamics simulation studies to examine the diffusional encounter dynamics in increasingly realistic models of the SOD—O_2^- system (Allison & McCammon, 1985; Allison, Ganti & McCammon, 1985*a*; Ganti, Allison & McCammon, 1985; Allison *et al.*, 1985*b*). In the first studies, SOD has been represented by a sphere of 2.85 nm radius. Two small, circular re-

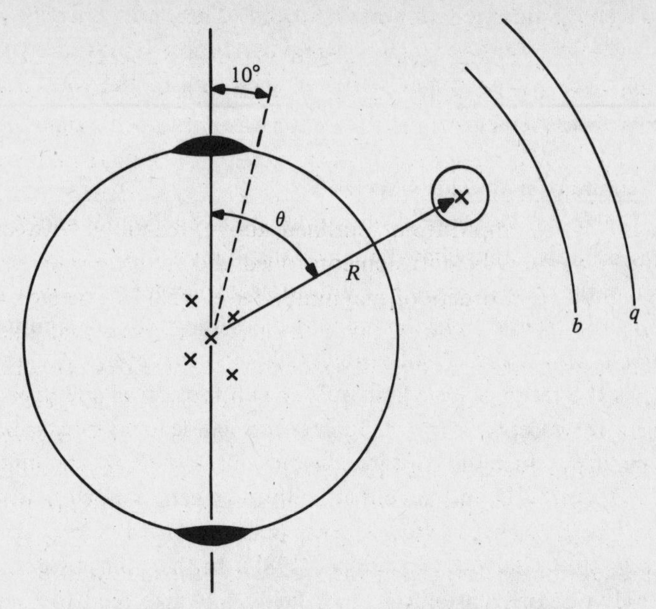

Fig. 8.1. Schematic illustration of a model used for simulating diffusion controlled reactions between the enzyme, superoxide dismutase, and its substrate, superoxide (Allison & McCammon, 1985). Crosses represent the positions of electrostatic charges. Active sites are indicated by the dark caps on the enzyme. Electrostatic interactions steer the diffusing substrates toward the active sites, leading to increased rates of reaction. (Copyright © 1985, American Chemical Society. Reprinted with permission.)

active patches (each subtending an azimuthal angle of 10°) were defined on opposite sides of the enzyme, as shown in figure 8.1. A cluster of five charges was placed in the enzyme to produce an electrostatic field that had monopole, dipole and quadrupole components outside the enzyme similar to those produced by all 76 ionic groups in the native structure. The O_2^- molecule was represented as a sphere of 0.15 nm radius with a single central charge. The dielectric constant was taken to be 78 throughout the system.

Diffusional encounters were simulated by computing trajectories of O_2^- in the field of the fixed enzyme (cf. section 4.10). This simple model system successfully reproduced the qualitative features of the SOD—O_2^- system. In particular, the rate constant was found to increase by nearly half when the noncentral components were added to the monopole field of the enzyme. Thus, the noncentrosymmetric field does in fact tend to steer substrate molecules into the active sites of the enzyme. Also, when salt effects were represented by a crude Debye–Hückel type model, it was observed that the reaction rate first increased, then decreased to a plateau as the solvent ionic strength was increased. The initial increase corresponds to the screening of the long range monopole repulsion, while the subsequent decrease corresponds to the screening of the shorter range, noncentral steering fields.

8.3 Nature of molecular association

The initial diffusional encounter between ligand and receptor molecules is more complicated than has commonly been assumed. As illustrated in the preceding section, intermolecular potentials can influence the rate of collisions between ligand molecules and receptor binding sites. In extreme cases, intermolecular potentials may capture any ligand that approaches the receptor and limit subsequent motion of the ligand to a region near the receptor surface. Such effects can lead to substantial rate enhancement, because the captured ligand will very likely encounter the binding site, even if its initial contact with the receptor was far from this site (Adam & Delbrück, 1968; Richter & Eigen, 1974; Chou & Zhou, 1982; Berg, 1984; Berg & von Hippel, 1985). In the case of ligands that bind in only one orientation, receptor fields may also tend to rotate the ligand into the correct orientation during the diffusional encounter step (Neumann, 1981; Margoliash & Bosshard, 1983; Matthew *et al.*, 1983; Allison, Srinivasan, McCammon & Northrup, 1984; Getzoff, 1985; Weber & Tollin, 1985). The randomizing effect of collisions with solvent molecules also tends to increase the rate of productive encounters beyond what might be expected on the basis of binding geometry (Berg & von Hippel, 1985). For example, a ligand that binds in only one orientation will have a relatively high probability of achieving that orientation during a given encounter if the ligand has a high rotational mobility.

The internal motions of a receptor may influence some details of its diffusional encounters with ligands (as a result of hydrodynamic coupling through the solvent, charge fluctuations, etc.), but more important kinetic effects are expected in the subsequent steps of the association process. The structural fluctuations of the receptor will generally include variations in the available volumes of its binding site, as well as in the disposition of charged, hydrogen bonding and other groups that are involved in ligand binding. The probability that a properly oriented ligand will actually be bound upon collision with the initial binding region at the receptor surface will therefore display a time dependence. The detailed character of the relevant binding site motions will depend on the particular ligand and receptor involved. For the initial entry of molecular oxygen into proteins such as myoglobin or hemoglobin, relatively small distortions of the protein matrix may be sufficient to open the necessary pathway (Frauenfelder *et al.*, 1979; Case & Karplus, 1979; Case & McCammon, 1986). A number of enzymes have mobile surface loops that must adopt certain conformations to allow ligand entry and then other conformations to secure the ligand in place. Examples include the 'flap' regions at the active sites of penicillopepsin (James & Sielecki, 1983) and triosephosphate isomerase (Banner *et al.*, 1975) and the activator binding loops in trypsin (Bode, 1979; Bennett & Huber, 1984). In yet other cases, hinge bending motions involving the relative displacement of large globular regions are required to open binding sites (McCammon *et al.*, 1976; Anderson, Zucker & Steitz, 1979; Newcomer *et al.*, 1981; Janin & Wodak, 1983; Bennett & Huber, 1984).

In all of the examples mentioned above, the receptor fluctuates among a variety of conformations, only some of which allow ligand binding. Thus, the receptor acts as a kind of gate that regulates access to the binding site. In the case of enzymes, such gating is often a necessary consequence of the functional design of the molecule. The enzyme must be able to open the binding site easily to allow substrate entry and product release, and yet be able temporarily to trap the substrate and bring the necessary catalytic groups into the correct positions around it. For other receptors, gating may be a consequence of the need to discriminate between different ligands or to activate the receptor for a subsequent functional step.

The kinetic effects of gate fluctuations can be estimated by considering the partially diffusion controlled reaction of particles whose intrinsic reactivity fluctuates with time (McCammon & Northrup, 1981; Northrup, Zarrin & McCammon, 1982*b*; Szabo, Shoup, Northrup & McCammon, 1982). In the simplest treatment, the reaction is described by the diffusion equation with a gated sink term

$$\frac{\partial \rho}{\partial t} = r^{-2}\frac{\partial}{\partial r}\left(r^2 D\frac{\partial \rho}{\partial r}\right) - k_s h(t)\rho\delta(r-R) \tag{8.2}$$

and a reflecting wall boundary condition at $r = R$. Here, $\rho(r,t)$ is the density of ligands at a distance r from the receptor at time t, D is the relative diffusion constant of ligand and receptor, k_s is the specific rate constant in the gate-open (reactive) state, R is the ligand–receptor contact distance, and $h(t)$ is a characteristic gating function that fluctuates between values of 0 (gate closed) and 1 (gate open). The average binding rate that would be observed in a conventional experiment is characterized by the bimolecular rate constant

$$k = \langle 4\pi R^2 k_s h(t)\rho(R,t)\rangle \tag{8.3}$$

where the brackets indicate a time average. The effects of gating on k are found to depend on such factors as the typical lifetimes of the gate-open and gate-closed states and the rate of motion of ligands relative to the receptor. In the limit of slow gate dynamics, k is just the rate constant for the gate-fixed-open case, multiplied by the fraction of the time that the gate is open. In other cases, substantial deviations from this intuitive result occur.

9

Recent developments and future directions

9.1 Introduction

Most of the work described in the preceding chapters has dealt with the development of theoretical methods for studying the dynamics of proteins and nucleic acids or with the application of these methods to characterize the general properties of such molecules. A number of research groups have also started to apply theoretical methods to show how dynamics are involved in the function of specific biological molecules. In the coming years, theoretical methods will assume increasing importance as tools for answering practical questions in biochemistry and pharmacology. These tools will be used in the systematic interpretation of experimental data and in the prediction of properties of molecules prior to the synthesis or experimental study of such molecules. Theoretical methods will also continue to be used to answer more fundamental questions concerning the dynamic nature of these molecules. This chapter describes some of the work that is underway along these lines, and some of the directions in which this work is likely to lead.

The strengths of the methods described in this book are a consequence of the fact that the models are so detailed – every atom can be treated, and the potential functions can represent all forms of interatomic interactions. At the same time, it is this very detailed representation that produces the most serious limitations on the methods. By today's standards, a 100 ps molecular dynamics simulation of a macromolecule with a molecular weight of 30000 in solution is computationally very expensive. As a consequence, important biological problems involving large molecules and motions on time scales longer than 1 ns cannot yet be easily treated. We are, however, optimistic that advances in theory and improved computational tools, both hardware and software, will render many of these problems tractable during the next few years.

In the following section, we outline a few of the methodological developments that will make theoretical studies of proteins and nucleic acids easier, more accurate, and more powerful. In the subsequent sections, we consider how theoretical tools can be applied to refine experimental results or make predictions concerning biomolecular structure and function. In the final section, we mention a few outstanding problems in the theoretical study of biomolecular dynamics.

9.2 Computing methods

There are basically two approaches to extending the time scales and the size of macromolecular systems that can be modeled on a digital computer. First, there continue to be steady increases in the speed and memory of computers and corresponding decreases in the cost of performing a given calculation. These improvements in computer hardware are briefly considered in section 9.2.1. Second, special techniques can be used to extend some of the methods described in this book. For example, molecular dynamics (which normally covers processes that occur on the subnanosecond time scale) can be extended to examine slow or rare processes whose characteristic times may be nanoseconds, microseconds, or even longer. Section 9.2.2 contains a brief survey of recent developments in methodology.

9.2.1 *Improvements in computer hardware*

Continuing developments in computer architecture are improving our capability to model biomolecular systems in several ways. First, superminicomputers such as the DEC VAX 11/780 with the speed and memory necessary to perform dynamical simulations of moderately large systems have become sufficiently affordable to be acquired by individual research groups. Second, attached processors such as the FPS 364 that are designed to perform arithmetic operations at rates significantly greater than the superminicomputers have become available at costs similar to the latter. Third, efforts are underway to develop specialized processors designed specifically to calculate nonbonded interactions or to do full macromolecular simulations (Hilhorst *et al.*, 1984; Barry & Crowther, 1985; Berendsen, 1985*c*; Levinthal, Fine & Dimmler, 1985; Ostlund & Whiteside, 1985). The potentially low cost of all of these machines means that they could be dedicated to particular problems for substantial periods of time. Even greater attention has focused on the increasing power and availability of supercomputers. These machines can be defined simply as the most powerful computers available at a given time. The high cost of these machines requires that they be shared among a large number of users, but the very great speeds and large memories of these computers may more than compensate for the limitations on access.

Clearly, the developments sketched above provide a spectrum of opportunity. At one end of this spectrum, one can perform a larger number of simulations of a given complexity in a given amount of real time. At the other end, one can simulate much more complicated systems than one could just a few years ago. One result is that the model systems used in simulation studies are becoming more detailed and realistic. For example, molecular dynamics simulations of proteins and nucleic acids with explicit hydrogen atoms in explicit solvent surroundings are replacing the previous simulations in which the hydrogen atoms and solvent were treated implicitly.

Although such comparisons rapidly become obsolete, it may be useful to provide estimates of the times required to carry out a dynamical simulation on several different machines. The reference system is the enzyme trypsin, plus the small inhibitor benzamidine, plus enough water molecules (4785) to approximate aqueous solution conditions; all hydrogen atoms that can participate in hydrogen bonds are explicitly represented (Wong & McCammon, 1986). This is a large system, comprising 16400 atoms. With the best available versions of the GROMOS molecular dynamics program (W. F. van Gunsteren *et al.*, personal communication), a 10 ps simulation of this system would require approximately 20 CPU hours on the CYBER 205 supercomputer, 250 CPU hours on the FPS 364 processor, and 2800 CPU hours on the VAX 11/780. To obtain reasonable speeds on machines such as the CYBER 205 (Berendsen, van Gunsteren & Postma, 1984; Gallion, Levy, Weiner & Hirata, 1986) or FPS 364 (T. P. Lybrand, personal communication) it is necessary to organize the computer program so that many arithmetic operations can be executed simultaneously. Similar considerations apply in the case of Brownian dynamics programs (Ganti & McCammon, 1986).

It must be borne in mind that serious reprogramming efforts may be required to take advantage of the special architecture of array processors, multiprocessors and supercomputers (Hockney & Jesshope, 1983; Bell, 1985). The speed of these machines generally results from their ability to carry out several computational tasks at once, and not all computer algorithms lend themselves to the parallel processing approach. As a general rule, the more recently a particular computer architecture has been developed, the more effort is required to convert pre-existing programs to a form that will (1) run at all and (2) run at maximum speed on the new machine. This is because the development of powerful preprocessing algorithms and compilers that can automatically optimize existing programs to run on a new computer usually lags several years behind the release of the machine. The detailed considerations of these reprogramming problems are beyond the scope of this book, but investigators who are considering a new computer would be well advised to consult someone who has already had experience in converting similar programs to that specific machine before making a commitment.

9.2.2 *Advances in methodology*

There are a variety of modifications that can be made to conventional simulation methods to extend the size of the systems that can be treated and the time scales that can be covered. The method of choice will, of course, depend on the particular problems. Several modified methods have already been described in chapters 4–8. Here we briefly review several newer developments in methodology that have not yet been applied very extensively.

In considering the size of the system being modeled, it is often the case that one is not truly interested in the detailed dynamics of the entire system, and only a small region (such as an active site or a ligand binding site) is of critical interest. In such a case, one can carry out full molecular dynamics on a small volume containing the region of interest if the volume is surrounded by a suitable boundary region in which atomic motions are modeled by stochastic dynamics (Berkowitz & McCammon, 1982; Brooks *et al.*, 1985). This procedure is described in more detail in section 4.7. If the size of the volume must be changed during the simulation, for example when the macromolecule is to be deformed, this can be accomplished by the use of suitable time dependent boundary conditions. The parameters of the boundary conditions may fluctuate spontaneously, as in molecular dynamics at constant pressure (appendix 3), or they may be changed systematically. The latter case has been shown to be particularly useful in the stretching deformations of DNA for the formation of an intercalation site (Prabhakaran & Harvey, 1985).

A second approach for extending the size and time scale of the simulations is to eliminate some details of the model. For example, the potential of mean force that acts on each residue of a protein can be determined in an ordinary molecular dynamics simulation, and this can then be used for the potential function of a simplified model of the molecule in which each residue is represented by a single body. If the potential of mean force is approximated by a quadratic function (the quasiharmonic approximation), then a normal mode analysis on this simplified model provides a relatively rapid method for studying the motions of the protein (Levy *et al.*, 1984*c*). Another example of this kind of model simplification is the method of Brownian dynamics (section 4.10), in which the equations of motion include terms for both stochastic collisions and viscous damping effects.

A third approach to extending effective time scales using a modified Monte Carlo procedure has recently been proposed by Noguti & Gō (1985). It was shown several years ago that traditional Monte Carlo methods are much less efficient than molecular dynamics for sampling macromolecular configurational space (Northrup & McCammon, 1980*a*). An improvement in the efficiency of the Monte Carlo method can be obtained by following the observation of Levy *et al.* (1984*b*) that the normal mode eigenvectors are the appropriate independent coordinates for the Monte Carlo steps. In the procedure developed by Noguti & Gō (1985), the step sizes along the normal mode eigenvectors are scaled so that the system takes the largest steps along the lowest frequency eigenvectors (in the directions of the softest normal modes). After several thousand Monte Carlo steps, a normal mode analysis is carried out on the current structure, which is not at an energy minimum, and the simulation proceeds from there. New normal mode analyses are similarly utilized after each 10^4 Monte Carlo steps. As would be expected, since the system is not at a minimum energy configuration, each normal mode analysis produces many negative eigenvalues. (About 15% of the frequencies are imaginary.) These authors treat these imaginary modes as if they were real ones by simply changing the signs of the eigenvalues, and they find that the model structure evolves in a reasonable and stable manner. The potential power of the method lies in the fact that the effective time step is reported to about 50 times larger than in molecular dynamics, although the potential gain in computational efficiency is substantially offset by the computationally expensive normal mode analysis at regular intervals. Also, the method provides no information on dynamics *per se*, but is rather a sophisticated tool for configurational sampling.

9.3 Biomolecular structure

Theoretical studies of motion in biological molecules have great potential utility in the refinement and prediction of molecular structure. In refinement, one might use the results of a molecular dynamics simulation to improve the accuracy of a model constructed from X-ray diffraction or NMR data, while in prediction one might use simulations to propose possible structures resulting from the insertion, deletion or modification of residues in a molecule of known structure. In this section, we describe recent work along these lines and indicate some possible extensions of this work.

X-ray diffraction studies of macromolecules provide fewer data than are required for a complete determination of the distribution of molecular structures within a crystal. This problem is circumvented in practice by two types of approximation. First, the molecular distribution is characterized by its lowest moments. For example, the model used to fit the data may comprise the mean positions of the atoms and the orientationally averaged mean square displacement of each atom from its mean position. In this case, one has four parameters to be fit for each atom, namely, three coordinates to define the mean position and a temperature factor to define the isotropic mean square displacement (cf. section 5.2.4). Second, the number of independent parameters may be reduced by requiring that these satisfy certain correlations or restraints. These restraints are typically based on the expected stereochemistry (e.g., bond lengths may be required to be close to certain ideal values) and on dynamical assumptions (e.g., the temperature factors of bonded atoms may be required to be of similar magnitude).

Quite recently, several groups have carried out model studies designed to improve the assumptions that are built into refinement algorithms. In these studies, X-ray diffraction data are first generated synthetically from the atomic position distributions observed in a molecular dynamics simulation. These data are then processed by the refinement algorithm, and the results (e.g., mean positions and the temperature factors of the atoms) are compared with the 'true' results obtained directly from the simulation. The assumptions built into the refinement algorithm are varied to find those assumptions that yield the best agreement with the simulation results. The studies have been based on different macromolecules, including the proteins BPTI (Yu, Karplus & Hendrickson, 1985) and myoglobin (Kuriyan, Petsko, Levy & Karplus, 1986), and a Z-DNA hexamer (Westhof *et al.*, 1986). These studies have shown that the assumption that each atom moves harmonically about a single mean position leads to small but significant errors in the positions of the most mobile atoms, and to underestimates of the amplitudes of motion of these atoms. These errors frequently arise when an atom hops back and forth among two or more stable positions; the probability density for the atomic position will then have two or more peaks, and the refinement algorithm typically tries to fit only one of these. Another finding is that improved estimates of the amplitudes of atomic motion can be obtained from existing restrained refinement algorithms by assuming that the correlation of temperature factors for bonded atoms is weaker than has commonly been supposed.

In addition to improving the assumptions that are built into refinement algorithms, it should in principle be possible to use simulations directly in the refinement of X-ray data. For example, improved agreement between models and crystallographic data can be obtained by subjecting partially refined structures to low temperature (about 150 K) molecular dynamics annealing (Westhof *et al.*, 1986). In general, however, until potential functions and simulation methods have been sufficiently improved to yield accurate mean crystal structures, simulations will probably be most useful in providing atomic position distribution functions that can be used to model the fluctuations about the experimentally observed positions. Both molecular dynamics (van Gunsteren *et al.*, 1983) and normal mode calculations (Levitt *et al.*, 1985) are likely to be useful in such work.

Molecular dynamics simulations also have considerable potential as a refinement tool in NMR determinations of protein and nucleic acid structures. A pioneering study in this area has been described by Kaptein *et al.* (1985). In this work, a set of 159 internuclear distances obtained by nuclear cross-relaxation measurements (the nuclear Overhauser effect) was used to construct and refine a model of the DNA binding domain of the lac repressor protein. The initial model comprised three alpha helices whose locations and relative orientation could be deduced from the internuclear distances. The model was refined by carrying out a molecular dynamics simulation subject to internuclear distance constraints corresponding to the NMR results. It was found that molecular dynamics was far more effective than energy minimization in eliminating the distance constraint violations present in the initial model. Subsequently, a similar approach has been used to study another protein fragment (Clore, Gronenborn, Brünger & Karplus, 1985). In principle, it should be possible to improve such calculations by using modified potential functions that qualitatively reflect solvation effects (cf. section 4.2).

Another method for building and refining models subject to constraints such as internuclear distances from nuclear magnetic resonance experiments is that due to Levitt (1983*c*). In this approach, a very powerful energy minimization algorithm (the variable metric method, with the protein single bond torsion angles ϕ, ψ and χ as the independent variables) is used to fold extended models into trial compact configurations that satisfy the constraints; these configurations are then further refined by annealing with molecular dynamics followed by conventional energy minimization. One novel feature of Levitt's treatment is the use of 'soft' atoms during the initial folding. These soft atoms are created by replacing the conventional van der Waals potential, where the interatomic forces are extremely strong at distances below about 0.25 nm, with a potential that permits atoms to pass through one another. This allows the generation of models with different chain threadings, providing a wide range of trial configurations suitable for examination and further refinement. Although the initial study was done on a model system, the method was shown to be suitable for building models subject to distance constraints (Levitt, 1983*c*). Consequently, it is an alternative approach to that of Kaptein *et al.* (1985) for using data from nuclear Overhauser measurements to determine macromolecular structure.

The spin–spin coupling constants obtained in NMR studies of macromolecules in solution should also be helpful in determining the structures of these molecules. Hoch, Dobson & Karplus (1985) have analyzed molecular dynamics simulations of BPTI and lysozyme to assess the effects of segmental motion in such determinations. As was done in some of the X-ray work described above, the simulation was used to generate synthetic experimental data (time averages of the instantaneous coupling constants), which were then compared with the experimental results expected for the average structures. Good agreement was found for the protein backbones, but the widespread occurrence of rotational isomerization leads to somewhat poorer agreement for the sidechains.

The useful range of computer simulation methods in the prediction of biomolecular structure is not yet clear. On the one hand, prediction of the tertiary structure of a protein given its sequence and a suitable potential function is well known to founder on the multiple minimum problem. On the other hand, prediction of the structure of a single site mutant protein given the parent structure is sometimes possible by systematic variation of the dihedral angles of the substituted amino acid and energy refinement of any low energy structures that are found (Shih, Brady & Karplus, 1985). Intermediate cases, for example, prediction of the structure of a protein with less extensive homology to a parent of known structure, will generally require the sampling of many low energy structures that differ in the conformations of many residues and are likely to involve formidable difficulties (Read, Brayer, Jurášek & James, 1984). In these cases, performing lengthy molecular dynamics simulations at normal or elevated temperatures and quenching a subset of the resulting configurations by energy minimization may allow one to generate low energy structures outside the region of the original local energy minimum (Levitt, 1983*c*; Di Nola *et al.*, 1984). Such procedures have been of practical value in suggesting alternative structures of peptides of possible medicinal interest (Hagler, Osguthorpe, Dauber-Osguthorpe & Hempel, 1985). Here and in the NMR structure refinement case, it will be necessary to study a wider range of systems to establish the radius of convergence of these sampling methods.

9.4 Biomolecular function

9.4.1 *Ligand binding*

Simulation studies are helping to clarify the nature of ligand binding in several biomolecular systems. The pathways of diffusional encounter between the substrate superoxide and enzyme superoxide dismutase have been explored by Brownian dynamics simulations, as described in section 8.2. In principle, such studies could be useful in the design of modified forms of this enzyme which have improved electrostatic steering fields for substrate molecules and correspondingly higher catalytic rates.

Theoretical studies of the atomic dynamics in complexes of carbon monoxide or dioxygen with hemoglobin or myoglobin are currently being pursued by several different groups. These studies attempt to go beyond previous work on such complexes (e.g., section 6.2) by considering the detailed dynamics of the protein as well as that of the ligand. Such studies should lead to more detailed interpretations of experimental kinetic data (Henry, Sommer, Hofrichter & Eaton, 1983; Ansari *et al.*, 1985; Gibson, Olson, McKinnie & Rohlfs, 1986) and more accurate parameters for theoretical studies based on phenomenological models (Agmon & Hopfield, 1983; Hanggi, 1983; Young & Bowne, 1984). Henry, Levitt & Eaton (1985) have examined the initial events that occur upon photodissociation of carbon monoxide from an isolated alpha chain of hemoglobin. Photodissociation was modeled by an abrupt change of potential function during dynamic simulations of the liganded complex; the potential was changed from one with an iron–ligand bond and forces favoring a flat heme to one with iron–ligand repulsion and forces favoring the nonplanar heme expected for the unliganded molecule. It was observed that heme moved to its nonplanar conformation with a half time less than 150 fs, which supports the interpretation of recent laser pulse spectroscopy experiments. Henry *et al.* (1985) plan to extend such simulations to study the response of the protein to the heme conformational change.

Case & McCammon (1986) have examined the entrance and exit of dioxygen from the heme pocket region in myoglobin by an approach similar to that used in studies of tyrosine ring rotations (section 6.2). In the first part of the calculation, the potential of mean force for ligand displacement at 300 K was determined along the primary path identified in earlier work by Case & Karplus (1979) (cf. section 6.2). A barrier of about 30 kJ/mol was found in the expected bottleneck region between the sidechains of His E7 and Val E11. The barrier is almost purely entropic in origin, which can be qualitatively rationalized in terms of the reduced translational and rotational mobility of the ligand in the bottleneck region compared to the heme pocket region. In the second part of the calculation, representative trajectories through the bottleneck region were computed. These trajectories seldom recross the transition state dividing plane, suggesting that the transmission coefficient for dioxygen hopping is larger than that for tyrosine flipping inside BPTI (section 6.2). This difference is likely due to the smaller size and correspondingly smaller collisional cross section of dioxygen compared to those of the tyrosine ring. Gating motions in the protein matrix are also evident.

New simulation approaches are also being developed to analyze or predict thermodynamic aspects of ligand binding. One of these, the thermodynamic cycle–perturbation method, has already been described in section 4.8. In this method, the relative free energy of binding ligands L and M to a receptor R is computed from two simulations in which L is transformed into M, first within the binding site of R and then in solvent surroundings. As described in section 4.8, the relative free energy of binding is equal to the difference in free energy of the 'nonphysical' transformations. This method has been used recently to calculate the relative free energy of binding differently substituted benzamidine inhibitors to the enzyme trypsin (McCammon, Karim, Lybrand & Wong, 1986*a*; Wong & McCammon, 1986*a*, *b*). For example, transformation of the para hydrogen atom to a fluorine atom yielded interaction free energy changes of 0.3 ± 2.0 kJ/mol and -3.5 ± 0.4 kJ/mol for inhibitors in the solvated enzyme and in liquid water, respectively. The binding of benzamidine is therefore predicted to be more favorable by 3.8 ± 2.4 kJ/mol than the binding of parafluorobenzamidine, and this difference is seen to be dominated by the difficulty of desolvating parafluorobenzamidine compared to benzamidine. The calculated difference in free energy of binding is consistent with experimental results.

The thermodynamic cycle–perturbation method and related approaches should be useful tools in the design of drugs, enzymes and other molecules. In comparisons of the binding of differently substituted ligands, the method can be regarded as a theoretical analogue of the Hansch method. The latter makes use of empirical correlations between binding and such properties of substituents as their relative solubility in different solvents to guide the design of bioactive molecules (Smith *et al.*, 1982; Carotti, Hansch, Mueller & Blaney, 1984). The theoretical approaches can also be used to predict the consequences of modifications in receptor structure as well as in ligand structure (see section 9.4.2). The capability of the method to deal accurately with solvation and entropy effects fills a gap in the existing theoretical methodology for molecular design (Beddell, 1984; Horn & De Ranter, 1984; Richards, 1984; Kollman, 1985).

9.4.2 *Enzyme activity*

Recently developed methods for the specific replacement of amino acid residues in enzymes by genetic manipulation have provided important new tools for enzymology (Fersht *et al.*, 1984). These methods can be used to study the roles of individual residues in substrate binding and catalysis (Craik *et al.*, 1985; Fersht *et al.*, 1985). Chemical techniques can also be used to modify the active sites of enzymes (Kaiser & Lawrence, 1984). In principle, these methods could be used to produce modified enzymes with activities and stabilities tailored for specific applications (Ulmer, 1983; Thomas, Russell & Fersht, 1985; Whitesides & Wong, 1985). It is likely that simulation methods will be useful in the analysis and prediction of the effects of such structural modifications.

In simple cases, the activity of an enzyme can be viewed as the product of two factors, namely the tendency of the enzyme to bind a given substrate and the rate of conversion of the bound substrate to products. For rat trypsin, replacement of Gly 216 by Ala reduces the activity of the enzyme toward arginine containing substrates primarily because of a 30-fold decrease in the substrate binding affinity (Craik *et al.*, 1985). Wong & McCammon (1986*a*,*b*) have carried out molecular dynamics simulations of bovine trypsin in water with and without a benzamidine inhibitor in the specificity pocket. Using the thermodynamic cycle–perturbation method (cf. section 4.8), they found that replacing Gly 216 by Ala resulted in about a tenfold decrease in affinity for this inhibitor. Because of the structural similarities between the rat and bovine enzymes, and between the guanidinium group of arginine and the amidinium group of the inhibitor, the similarity of the theoretical and experimental results is encouraging. Analysis of the components of the thermodynamic cycle suggests that steric effects are responsible for the reduced affinity of the mutant enzyme (figure 9.1). The thermodynamic cycle–perturbation method could also be used to predict the effects of mutations on the rate of reaction of bound substrate by calculation of the relative free energies of activation. Such theoretical studies are in a very early stage. Much work will have to be done to evaluate the potential and limitations of these methods. One factor that certainly must be considered is the effect of a mutation on the overall stability of the folded protein (Craik *et al.*, 1985).

Simulations are also likely to play an increasing role in descriptions of the detailed mechanisms of enzyme function. Quite recently, Chandrasekhar *et al.* (1985), Jorgensen *et al.* (1986) and Madura & Jorgensen (1986) have shown how high level quantum chemical calculations for reacting solute molecules and empirical potential functions for solute–solvent and solvent–solvent interactions can be combined with umbrella sampling

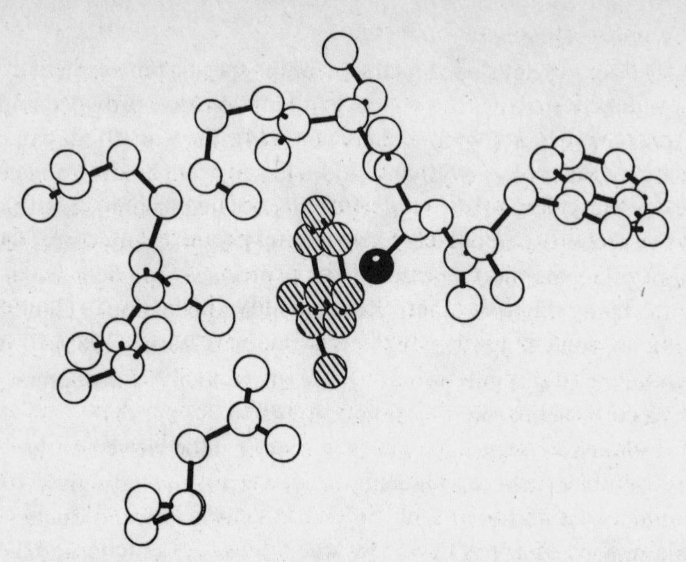

Fig. 9.1. Single site mutation in the trypsin specificity pocket, viewed from the surface of the enzyme. The methyl group (solid sphere) added in the Gly 216 → Ala 216 mutation projects toward the benzamidine inhibitor (hatched spheres). Steric conflict with this methyl group is predicted to reduce the affinity of the enzyme for the inhibitor. The coordinates are from one instantaneous configuration generated in a molecular dynamics study of the free energy of inhibitor binding (Wong & McCammon, 1986*a*, *b*).

techniques (section 4.8) to compute the free energy as a function of reaction coordinate for nucleophilic substitution and addition reactions of small molecules in liquid water. The development of these and parallel techniques for enzyme–substrate systems (Weiner, Seibel & Kollman, 1986) will open the way for rate constant calculations for enzymes based on transition state theory (cf. section 6.4). Such efforts should eventually permit the modification of empirical potential functions to allow rapid calculation of the energy and forces associated with covalent rearrangements in enzymes. These functions could be used in activated trajectory calculations to provide details concerning nonequilibrium contributions to rate constants, mechanisms of accumulation and disposal of energy between reaction steps, and so on. Exploratory studies of this kind using a simple model potential function have been described by Warshel (1984).

9.4.3 *Macromolecular association*

The flexibility of biopolymers plays an important role in the formation of complexes of these molecules. Here, we mention three representative examples in which theoretical studies are likely to be useful in the future.

A number of studies have shown that thermal mobility correlates strongly with antigenicity in proteins, as reviewed by Tainer *et al.* (1985) and by Berzofsky (1985). Two particular correlations relate to peptides that correspond to segments of whole proteins. First, antibodies to peptide fragments of a protein are relatively likely to bind to the native protein if the corresponding protein segment is highly mobile. Second, antibodies to native protein are relatively likely to bind peptide fragments if the latter correspond to mobile segments of the protein. As stressed in the above reviews, many factors other than mobility influence recognition and binding phenomena in the immune system. Nevertheless, the two types of correlation with protein mobility are of great interest. In practical terms, the two correlations should be helpful in the design of peptide vaccines and of affinity elements for chromatographic fractionation of antisera, respectively. In such applications, molecular dynamics simulations have potential as design tools; such simulations could be used to indicate the mobile regions of protein antigens without requiring temperature factor analyses and the associated corrections for environmental effects such as suppression of local mobility due to crystal packing. In more fundamental terms, the correlation of mobility and antigenicity is of interest with respect to the immune response. The flexibility of antigens suggests that different antigens can adopt structures complementary to a given antibody (at some cost in affinity due to associated strain energy and reduced conformational entropy); the immunologic repertoire corresponding to a given amount of genetic information would be thereby increased (Westhof *et al.*, 1984).

The associations of certain biopolymers involve hinge bending motions in one of the molecules. Examples such as enzymes and viral coat proteins have already been mentioned (cf. section 7.1). A particularly interesting case is the lactose repressor protein of *E. coli.* This molecule appears to have some structural features in common with the L-arabinose binding protein (Müller-Hill, 1983; Sams, Vyas, Quiocho & Matthews, 1984). As described in section 7.2, the latter protein undergoes a large scale hinge bending motion upon binding of arabinose, so that the sugar is buried within the protein. The repressor apparently undergoes a similar motion upon binding of the sugar lactose; in the closed conformation, the protein dissociates from the operator region of the lac operon, allowing transcription of the associated genes. A recent study of a mutant repressor has shed additional light on the importance of hinge bending in this system (Chakerian, Pfahl, Olson & Matthews, 1985). The mutation, replacement of alanine by valine at one site, appears to lock the repressor into the closed conformation. The rate constant for binding lactose is reduced by a factor of 150–300, probably due to infrequent opening of the gate to the binding site (cf. section 8.3). Even in the absence of lactose, the repressor has a low affinity for the operator because it is locked into the liganded conformation. Studies of other repressor molecules show that hinge bending motions are also involved in binding of these molecules to DNA (Little & Hill, 1985).

The flexibility of DNA plays a critical role in gene regulation and in packaging the genetic material in viruses and other assemblies. An excellent, brief account of this subject has been given recently by Widom (1985). One system of interest involves the catabolite gene activator protein (CAP) of *E. coli.* When cyclic AMP is attached to this protein, CAP binds to the DNA at a number of sites, where it stimulates binding of RNA polymerase and gene transcription. For a variety of reasons, summarized by Widom (1985), it appears that CAP does not interact directly with RNA polymerase, but rather distorts the DNA so as to facilitate the binding of the polymerase. Conceivably, simulation studies would be helpful in characterizing the detailed nature of the DNA distortion that is involved.

Simulation can also be expected to play an increasing role in studies of the diffusional stage of macromolecular association. Dickinson & Honary (1986) have, for example, used Brownian dynamics to characterize the time and space distribution of local reactions catalyzed on the surface of a colloidal particle by a diffusing enzyme. Their results help to explain the kinetics of aggregation of certain biocolloids such as milk proteins. There are also many associations leading to complex formation between biopolymers in which electrostatic steering, gated binding, or other complicating factors play a role (sections 2.4, 3.4 and 8.3). Brownian dynamics studies are likely to be helpful in the detailed description of such associations.

9.5 Outstanding problems

In this final section, we briefly mention a few areas in which the development or application of theoretical methods appears to present especially formidable difficulties.

The energy functions that form the basis for simulation studies of biopolymers contain a number of approximations, whose consequences depend on the specific application of interest (cf. section 4.2). Many questions remain concerning the exact domains of validity of the existing functions and how these functions can be improved. Considering the types of functions used in detailed energy minimization and molecular dynamics calculations for example, the need to explicitly include more than two body interactions (e.g., electronic polarizability) has often been discussed. Experience with studies of aqueous solutions of small molecules suggests that two body effective potentials are sufficient to answer many structural and thermodynamic questions, even ones that involve the introduction of ions into water (Jorgensen *et al.*, 1986; Lybrand *et al.*, 1986). It may, however, be necessary to include explicit polarization effects in studies of the mechanism of ionic association or other processes where solvent molecules will be subject to particularly large variations in electric field. Several groups have developed potential functions that include polarization effects (Lybrand & Kollman, 1985, and references therein), but these have not been extensively applied. For simulations of processes in which larger electronic rearrangements occur (e.g., formation or breaking of bonds, electron transfer), our knowledge of the potential surfaces necessary for biological studies is almost nonexistent (Frauenfelder & Wolynes, 1985; Jorgensen *et al.*, 1986).

As discussed in chapters 7 and 8, it is often desirable to avoid the explicit treatment of solvent by making use of an appropriate potential of mean force. Despite much work in this area (cf. section 4.2), there is as yet no generally practical technique for computing such potentials that includes all the important factors of biopolymer shape, biopolymer and solvent dielectric characteristics, and distribution of mobile salt ions. Among the more promising approaches over the long term are those that seek to extend to biopolymer solutions integral equation techniques that have proven useful in studies of simpler solutions (Chandler, 1982; Friedman, 1985*a*). Such methods were used some time ago to obtain an effective potential function for Brownian dynamics studies of a model protein sidechain in water (Levy *et al.*, 1979). Recent theoretical developments have increased the power of these techniques; they have now been used successfully to determine the potentials of mean force for conformational changes of a dipeptide in water (Pettitt & Karplus, 1985) and for a simple nucleophilic substitution reaction in water (Chiles & Rossky, 1984). Although much developmental work is necessary before these techniques will yield detailed results for macromolecular systems, the fact that they require orders of magnitude less computation than do full simulations imbues this approach with great promise.

Given the necessary potential functions, there still remain daunting problems in the study of certain types of phenomena. The mechanism of solvation changes associated with ligand binding or conformational transitions may be intrinsically difficult to determine in some cases. Such solvation changes may involve a large number of steps corresponding to rearrangements of the hydrogen bonds of individual solvent molecules (Karim & McCammon, 1986). If the solvent molecules are bound to charged groups or subject to steric constraints, the individual steps may have significant energies of activation. The activation requirements of the individual steps may lead to long characteristic times, and the number of steps may preclude use of the methods described in section 6.2. An extreme example would be the binding of a sugar molecule to the L-arabinose binding protein, in which solvent must be stripped from charged groups in the binding site as the protein sequesters the sugar in a hinge bending process (section 7.2). Analogous 'many step' problems arise in certain large scale structure changes of proteins (Ansari *et al.*, 1985).

As a final example, consider the difficulty of dealing with a single but complicated activated process. The method outlined in section 6.2 requires the definition of a reaction coordinate that explicitly recognizes each interaction that makes a systematic contribution to the free energy barrier for the process. As was already apparent in the discussion of tyrosine ring rotations, the elements that are important may not always be obvious even in simple cases. More extensive processes such as allosteric transitions or enzyme activation are thus not presently accessible to systematic study. Recent work by Nguyen & Case (1985) represents an important direction in the development of systematic techniques to determine reaction coordinates, but these techniques are still reliable only for systems of up to a few dozen atoms.

APPENDIX 1

Numerical integration of the equations of motion

We present here three methods for numerically approximating the Taylor series for the position, equations (4–18). Two of these are second order methods (the Verlet and Beeman algorithms) and one is fifth order (the Gear algorithm). In addition, we describe a modification to the Verlet method that is frequently used to reduce computer time by fixing some or all of the covalent bond lengths (the SHAKE algorithm).

A1.1 The Verlet method (Verlet, 1967)

If v is the average velocity during the time interval between t and $t+\Delta t$, then the position at the end of the interval is exactly

$$x(t+\Delta t) = x(t)+v\Delta t \qquad \text{(A1.1)}$$

We now assume that v is very nearly equal to the instantaneous velocity at the midpoint of the interval,

$$v = \dot{x}(t+\Delta t/2) \qquad \text{(A1.2)}$$

This quantity can be calculated from the velocity $\dot{x}\ (t-\Delta t/2)$ if we know a, the average acceleration during the interval from $t-\Delta t/2$ *to* $t+\Delta t/2$, by the exact relationship

$$\dot{x}(t+\Delta t/2)=\dot{x}(t-\Delta t/2)+a\Delta t \qquad \text{(A1.3)}$$

Assuming that a is very nearly equal to the instantaneous acceleration at the midpoint of this interval,

$$a = \ddot{x}(t) \qquad \text{(A1.4)}$$

we obtain for the new velocity

$$\dot{x}(t+\Delta t/2) = \dot{x}(t-\Delta t/2)+\ddot{x}(t)\Delta t \qquad \text{(A1.5)}$$

We can then update the position by combining equations (A1.1) and (A1.2):

$$x(t+\Delta t) = x(t)+\dot{x}(t+\Delta t/2)\,.\,\Delta t \tag{A1.6}$$

This procedure is frequently called the *leapfrog method*, because the velocity is calculated at odd half integral multiples of Δt (equation (A1.5)), while position is calculated at integral multiples of Δt (equation (A1.6)). From the updated position, a new acceleration is calculated, and the velocity can be updated (equation (A1.5)), starting the cycle anew.

If one is not interested in calculating the velocity, that step can be eliminated by combining equations (A1.1)–(A1.4), giving an expression for directly updating the position:

$$x(t+\Delta t) = 2x(t)-x(t-\Delta t)+\ddot{x}(t)\,.\,(\Delta t)^2 \tag{A1.7}$$

Some authors follow the velocity at time points corresponding to those at which the positions are determined, namely, at integral multiples of Δt, using the relationship

$$\dot{x}(t) = [x(t+\Delta t)-x(t-\Delta t)]/(2\Delta t) \tag{A1.8}$$

There are two disadvantages to using equations (A1.7) and (A1.8) rather than equations (A1.5) and (A1.6). First, the velocity is calculated as a difference, and substantial numerical errors can occur if this is a small difference between large numbers. This can lead to errors in the calculated kinetic energy and temperature. Second, the position in equation (A1.7) is not calculated from the velocity in equation (A1.8), so it is not possible to control the temperature or energy of the system by rescaling velocities (cf. appendix 2). Neither of these shortcomings occurs if equation (A1.5) and (A1.6) are used.

The principal advantages of the Verlet algorithm (equation (A1.7)) are its conceptual simplicity and the fact that it can be easily modified to give an algorithm with constraints on internal coordinates such as bond lengths and bond angles. These modifications are discussed in the section on the SHAKE algorithm, below.

A1.2. The Beeman method (Beeman, 1976; Levitt, 1983*a*; Levitt & Meirovitch, 1983)

This procedure assumes that, during the time interval Δt, the acceleration varies linearly with time according to the relationship

$$a(t_0+t) = a(t_0)+t[a(t_0)-a(t_0-\Delta t)]/\Delta t.$$

If this expression is integrated, we obtain

$$v(t_0+t) = v(t_0)+ta(t_0)+t^2[a(t_0)-a(t_0-\Delta t)]/2\Delta t$$

Integrating again gives

$$x(t_0+t) = x(t_0)+tv(t_0)+t^2a(t_0)/2+t^3[a(t_0)-a(t_0-\Delta t)]/6\Delta t$$

If we wish to take a step in time, we set $t = \Delta t$, which gives

$$x(t_0+\Delta t) = x(t_0)+v(t_0).\Delta t+[4a(t_0)-a(t_0-\Delta t)].(\Delta t)^2/6$$

This can be rewritten in a form consistent with the notation of the previous section by dropping the subscript, so that the reference time point is simply t rather than t_0:

$$x(t+\Delta t) = x(t)+v(t).\Delta t+[4a(t)-a(t-\Delta t)].(\Delta t)^2/6 \qquad \text{(A1.9)}$$

If we solve this equation for $v(t)$ and use equation (A1.7) to eliminate $x(t+\Delta t)$, we obtain, on rearranging,

$$v(t) = [x(t)-x(t-\Delta t)]/\Delta t+[2a(t)+a(t-\Delta t)].\Delta t/6$$

This is equivalent to

$$v(t+\Delta t) = [x(t+\Delta t)-x(t)]/\Delta t+[2a(t+\Delta t)+a(t)].\Delta t/6$$

The first bracketed term on the right-hand side of this expression can be determined from equation (A1.9), giving the final expression for the updated velocity,

$$v(t+\Delta t) = v(t)+[2a(t+\Delta t)+5a(t)-a(t-\Delta t)].\Delta t/6 \qquad \text{(A1.10)}$$

Equations (A1.9) and (A1.10) provide a direct method for stepping the system through time. The position is updated first, so that the acceleration corresponding the new position, $a(t+\Delta t)$ can be calculated, since this quantity is required to update the velocity (equation (A1.10)).

The principal advantage of this method for molecular dynamics lies in its efficiency. Since it treats the first-order (linear) variation of acceleration during the time interval, it is effectively a third-order method. Consequently, larger time steps can be taken than with the Verlet algorithms, while still conserving energy within tolerable limits, giving an increase in efficiency. Levitt and Meirovitch (1983) report that this method is between two and four times as efficient as the Gear algorithm, and about 20% more efficient than the SHAKE algorithm when used under identical conditions. However, it does require more memory than the Verlet algorithm, which may be a drawback in certain applications.

A1.3 The fifth-order Gear method (Gear, 1971; McCammon *et al.*, 1977; 1979)

To include information over several time steps in the integration scheme, higher order derivatives can be included in the truncated Taylor series. The procedure described here includes terms through the fifth derivative. If we write

$$X_n = n!(\Delta t)^n(\mathrm{d}^n x/dt^n)$$

and if we now use unprimed quantities to represent values at time t and primed quantities to represent the values predicted at time $t+\Delta t$ by the truncated Taylor series, we obtain

$$X_0' = X_0+X_1+X_2+X_3+X_4+X_5$$

$$X_1' = X_1+2X_2+3X_3+4X_4+5X_5$$

$$X_2' = X_2+3X_3+6X_4+10X_5$$

$$X_3' = X_3+4X_4+10X_5$$

$$X_4' = X_4+5X_5$$

$$X_5' = X_5$$

To compensate for truncation errors, the force that corresponds to the configuration X' can be used to correct the values predicted by the last six equations. If we denote that force by F' and write

$$A = F'(\Delta t)^2/2-X_2'$$

then the corrected values for the new position and derivatives are

$$X_0(t+\Delta t) = X_0'+(3/16)A$$

$$X_1(t+\Delta t) = X_1'+(251/360)A$$

$$X_2(t+\Delta t) = X_2'+A$$

$$X_3(t+\Delta t) = X_3'+(11/18)A$$

$$X_4(t+\Delta t) = X_4'+(1/6)A$$

$$X_5(t+\Delta t) = X_5'+(1/60)A$$

A1.4 The constrained Verlet (SHAKE) method (Ryckaert *et al.*, 1977; van Gunsteren & Berendsen, 1977)

Among the most serious limitations of molecular dynamics simulations on large systems is the fact that they are computationally expensive. Consequently, methods that can reduce the required CPU time are much to be desired.

One approach to reducing computer time is to increase the time step Δt. The maximum time step is determined by the requirement that Δt be small by comparison to the period of the highest frequency motion in the system being simulated. Since the highest frequency motions in a macromolecular model are the bond stretching vibrations, different algorithms that can constrain the bonds to a fixed length have been investigated. Here we briefly describe the most popular of these algorithms, SHAKE.

We begin with the assumption that all constraints can be represented by fixed interatomic distances. (This is obviously true for constrained bond lengths. Valence angle constraints can also be expressed in this way, since a trio of atoms x, y, and z can be held in a rigid geometry by three constraints, fixing (1) bond xy, (2) bond yz, and (3) either the angle xyz or the distance xz.)

If the kth constraint is on the distance between atoms i and j, then it can be expressed by the relation

$$\mathbf{r}_{ij}^2 - d_{ij}^2 = 0 \tag{A1.11}$$

where $\mathbf{r}_{ij}$ is the vector from atom i to j:

$$\mathbf{r}_{ij} = \mathbf{r}_j - \mathbf{r}_i$$

In a numerical simulation, equation (A1.11) is not generally exactly true, so we say that the constraint is satisfied whenever the relative deviation is sufficiently small, i.e.

$$s_k = (\mathbf{r}_{ij}^2 - d_k^2)/d_k^2 < \varepsilon \tag{A1.12}$$

where ε is a specified tolerance. SHAKE is an iterative procedure that adjusts the atomic positions after each time step in order to simultaneously satisfy all the constraints, iterating until s_k is smaller than ε for all values of k. Note that this could be done exactly, but the exact approach would require the solution of a quadratically nonlinear set of matrix equations.

If we start with a structure where all constraints are perfectly satisfied, then, for all n constraints ($k = 1, 2, \ldots, n$), the n equations (A1.11) are satisfied. If a normal, unconstrained Verlet step is then taken, the resulting positions of atoms i and j, indicated by the vectors $\mathbf{r}_i^*$ and $\mathbf{r}_j^*$, will generally be such that equation (A1.11) no longer holds. We need to find appropriate adjustments to the atomic positions, denoted $\delta\mathbf{r}_i$ and $\delta\mathbf{r}_j$, so that the adjusted positions,

$$\mathbf{r}_i'' = \mathbf{r}_i^* + \delta\mathbf{r}_i$$

$$\mathbf{r}_j'' = \mathbf{r}_j^* + \delta\mathbf{r}_j$$

are such that all n constraints are again satisfied. It can be shown that the adjustments are of the form

$$\delta \mathbf{r}_i = -\sum_j g_{ij}\mathbf{r}_{ij}/m_i \tag{A1.13}$$

where m_i is the mass of atom i, $g_{ij} = g_{ji}$ for all constrained pairs of atoms, and $g_{ij} = 0$ if the distance $\mathbf{r}_{ij}$ is not constrained. The problem is the determination of g_{ij}.

Formally, this is a problem involving holonomic constraints, and the Lagrangian formulation of classical mechanics (Goldstein, 1980) is a logical approach. For very small molecules or for very simple molecules, like n-alkanes, the Lagrange approach can be used, for g_{ij} is simply related to the Lagrange multiplier for the kth restraint. Ryckaert *et al.* (1977) give a full treatment of this problem for an unbranched semirigid chain of arbitrary length. For large molecules, however, it is difficult to find the correct set of generalized coordinates, and the Lagrange approach is inappropriate.

SHAKE is an alternative method that considers the constraints in succession, treating each g_{ij} as an unknown that can be calculated. If we consider only the kth constraint, which fixes the distance between atoms i and j, the adjustments to the positions of each of these atoms is, by equation (A1.13),

$$\delta^k \mathbf{r}_i = -g_{ij}\mathbf{r}_{ij}/m_i \tag{A1.14}$$

$$\delta^k \mathbf{r}_j = g_{ij}\mathbf{r}_{ij}/m_j \tag{A1.15}$$

where we have used the superscript as a label for this constraint. At the step where the kth constraint is to be applied, the position of each atom has already been adjusted for the previous (k-1) constraints. Remembering that we have written $\mathbf{r}_i^*$ and $\mathbf{r}_j^*$ for the atomic positions after the unconstrained Verlet step, we note that as we prepare to apply the kth constraint, the interatomic vector is thus given by

$$\mathbf{r}' = (\mathbf{r}_j^* + \sum_{k'<k} \delta^{k'}\mathbf{r}_j) - (\mathbf{r}_i^* + \sum_{k'<k} \delta^{k'}\mathbf{r}_i)$$

In this last equation, we have dropped all subscripts on the left-hand side, because we are referring to atoms i and j and constraint k from this point on.

Similarly, we can drop subscripts and use the vector $\mathbf{r}$ to denote the interatomic vector $\mathbf{r}_{ij}$ for the starting structure, prior to taking the unconstrained Verlet step. (Recall that $\mathbf{r}$ did satisfy the constraint condition, equation (A1.11).) We also introduce the following definitions to simplify the notation:

$$\delta\mathbf{r} = \delta^k\mathbf{r}_j - \delta^k\mathbf{r}_i$$

$$g = g_{ij}$$

$$d = d_{ij}$$

$$M = 1/m_i + 1/m_j$$

The requirement that the final atomic positions are such that the kth constraint is satisfied can be expressed by

$$(\mathbf{r}'')^2 - d^2 = 0$$

where

$$\mathbf{r}'' = \mathbf{r}' + \delta\mathbf{r}$$

These last two equations give

$$(\mathbf{r}' + \delta\mathbf{r})^2 - d^2 = 0$$

Inserting equations (A1.14) and (A1.15) into the definition for $\delta\mathbf{r}$ gives

$$\delta\mathbf{r} = Mg\mathbf{r}$$

Combining these last two equations, we obtain

$$M^2g^2\mathbf{r}\,.\,\mathbf{r} + 2Mg\mathbf{r}\,.\,\mathbf{r}' + \mathbf{r}'\,.\,\mathbf{r}' - d^2 = 0 \tag{A1.16}$$

This quadratic equation for g can be solved iteratively by noting that $Mg \ll 1$ (because $\delta\mathbf{r} \ll \mathbf{r}$ for sufficiently small Verlet steps), so the first term can be dropped for computational efficiency, giving

$$g = (d^2 - \mathbf{r}'\,.\,\mathbf{r}')/(2M\mathbf{r}\,.\,\mathbf{r}') \tag{A1.17}$$

We note that g is proportional to the amount by which the constraint is distorted from its ideal length, as would be expected, and that $g < 0$ when the bond ij is stretched. Equations (A1.14) and (A1.15) show that to restore the constraint in this case, atoms i and j are moved toward one another by amounts that are inversely proportional to their respective masses. The vector relationships of this procedure are shown in figure A1.1.

The satisfaction of the kth constraint will generally destroy some of the previously satisfied constraints ($k' < k$). The procedure is repeated over all constraints with a sufficient number of iterations so that all constraints are adequately satisfied, i.e. until $s_k < \varepsilon$ for all values of k (Eq. A1.12). This method simultaneously assures that the quadratic equation (A1.16) is also

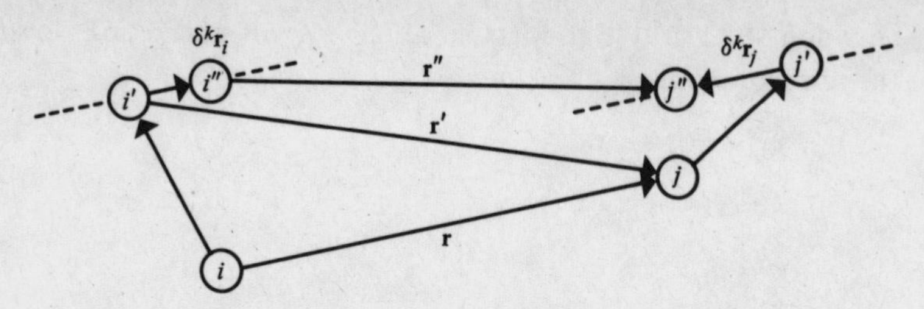

Fig. A1.1. Vector relationships in SHAKE (see text for discussion).

solved within the specified tolerance, because successive values of g from the linear approximation (equation (A1.17)) are smaller and smaller.

The effects of the size of the tolerance ε on computational efficiency have been explored by Ryckaert *et al.* (1977) and by van Gunsteren & Karplus (1982*a*); typical values for simulations on macromolecules are on the order of $\varepsilon = 10^{-4}$. It has been found that using SHAKE to constrain bond lengths increases computational efficiency over the unconstrained Verlet method by a factor of three, and that when bonds are constrained, the dynamical properties of macromolecules are not significantly affected on time scales longer than about 0.05 ps. Constraints on the valence bond angles are not recommended except for those in water molecules (van Gunsteren & Berendsen, 1977; van Gunsteren & Karplus, 1982*a*).

SHAKE can be applied to the Gear algorithm as well, and the dynamic results are similar to those with the Verlet/SHAKE (van Gunsteren & Berendsen, 1977). The Gear/SHAKE algorithm is rarely used however, because it is computationally more complex and expensive than the Verlet/SHAKE algorithm.

APPENDIX 2

Detailed description of computer programs and procedures for energy minimization and molecular dynamics

This appendix is intended to serve as a resource to which one can turn when beginning to design, use or modify a package of computer programs for molecular modeling studies using energy minimization, normal mode analysis, and/or molecular dynamics. Among the more common packages (and the principal references on each) are AMBER, developed at the University of California at San Francisco (Weiner & Kollman, 1981); CEDAR, developed at the University of North Carolina (Carson & Hermans, 1985); CHARMM, developed at Harvard University (Brooks *et al.*, 1983); and GROMOS, developed at the University of Groningen (van Gunsteren *et al.*, 1983, and references therein).

Topics to be discussed here include the data files required for simulations and the sequence of operations necessary to properly initiate a molecular dynamics simulation. As an example, we describe one particular package of programs, those developed for our studies on the structure and dynamics of phenylalanine transfer RNA (Harvey & McCammon, 1981; Tung *et al.*, 1984; Harvey *et al.*, 1984, 1985; Prabhakaran *et al.*, 1983, 1985*a*). We hope that a discussion of this example will clarify many of the details of the computer algorithms and answer most of the questions commonly posed by newcomers to the field. We do not discuss the various numerical methods that are used, since they are described elsewhere (see section 4.4 on energy minimization and appendix 1 on molecular dynamics).

A2.1 Data files

The common feature of programs for energy minimization and molecular dynamics is the potential energy function (equation (4.1)). Using that expression, the computer calculates the potential energy of any given conformation of the molecule. If the forces on the atoms are required, they can be calculated from the derivatives of the potential energy function with respect to the appropriate coordinates. In that context, it is useful to distinguish between the *Cartesian coordinates* and the *internal coordinates*. The former represent the positions of the atoms in real space and are normally given as a triplet of numbers for each atom (x, y, z). The latter consist of numbers describing the molecular structure in terms that are useful for calculating energies or manipulating the molecule in ways that consider the covalent structure; they include bond lengths, covalent bond angles, and so on.

The following is a brief description of the data files that are required for modeling molecular structure and dynamics with atomic resolution. The reader who is using a package of programs should be aware that the terminology for naming the files and the ways in which they are grouped together will depend on which particular package is being used.

The *atom data file* contains information on each of the atoms in the molecule. The atoms are numbered sequentially, usually following the standard numbering scheme of the International Union for Pure and Applied Chemistry. (For references on the IUPAC conventions, see the *Instruction for Authors* of any of the principal biochemical journals.) This file contains the name of each atom and, for proteins and nucleic acids, identifies the residue in which the atom is located. It also specifies the partial charge, mass and the type of each atom. The designation of atom type makes it possible to properly represent the stereochemical properties of each atom. It is necessary, for instance, to distinguish between carbon atoms with sp^3 and sp^2 hybridizations, so that carbon–carbon single and double bonds are given their proper bond strengths and ideal lengths, and to provide the correct geometry for tetrahedral and trigonal bond angles. Since about half the atoms in a macromolecule are hydrogens, and since the positions of hydrogens are not determined in crystallographic studies, it is common to reduce the computational burden by omitting most or all of the hydrogen atoms from these models. Their effects are approximated by using an 'extended atom' for each heavy atom plus the hydrogens bonded to it. Extended atoms are given suitable properties to implicitly include the effects of the hydrogen atoms. For example, the van der Waals radius of the extended atom type that represents a methyl group will be substantially larger than that of the atom type for a carbon atom whose substituents do not include any hydrogens. In many studies, the only hydrogens that are treated explicitly are the 'polar hydrogens' found on hydrogen bond donors. This allows the effects of the hydrogen bond geometry, such as the donor–hydrogen–acceptor angle, to be properly treated.

The *coordinate file* contains the Cartesian coordinates of all of the atoms of the model. The starting structures for many studies on macromolecules will be the coordinate sets determined by X-ray diffraction, over 250 of which are currently available from the Brookhaven Protein Data Bank (Bernstein *et al.*, 1977; Abola, Bernstein & Koetzle, 1985). Other investigations will begin with model coordinates, such as those for various conformations of nucleic acids that were derived from fiber diffraction studies (Arnott *et al.*, 1976), or coordinates generated by other modeling studies.

The *molecular topology file* describes the covalent structure of the molecule in terms such that the computer can calculate values of the internal coordinates. For example, a list of which pairs of atoms are covalently bonded to one another is necessary for the determination of bond lengths. The topology file gives the total number of bonds and, for each bond, it specifies the two atoms in the bond by giving their identification numbers, e.g. (I, J), and by specifying the type of bond. The bond type is determined by the atom types for atoms I and J, allowing discrimination between carbon–carbon single and double bonds, for example. Similarly, bond angles are specified by listing triplets of atoms (I, J, K), and giving the type of bond angle for each. The bond angle type will, of course, depend on the atom types for the three atoms forming the angle. Torsional rotations about the bond between atoms J and K will be identified by the quartet (I, J, K, L), and a torsional angle type will also be given for each torsion. Frequently, 'improper torsions' are also listed. These are used, for example, wherever one has tetrahedral geometry about atoms where a substituent atom has been incorporated into the central extended atom, because it is necessary to prevent accidental chiral inversions at these centers. Improper torsions may also be used to maintain coplanarity of groups of atoms. Normally, the potential function for improper torsions is quadratic, and a large force constant is used. If atom 1 is the central atom and its tetrahedral substituents are atoms 2, 3, 4, and a hydrogen atom that does not appear explicitly, the improper torsion angle 1–2–3–4 will be about $\pm 35.3°$, with the choice of sign depending on the handedness of the chiral center.

The foregoing components of the topology file are essentially immutable, because normally the making or breaking of covalent bonds is forbidden during the kinds of simulations described in this book. In addition to the foregoing topological elements, most algorithms allow the definition of explicit hydrogen bonds by specifying donor–acceptor pairs, (I, J). The type of each hydrogen bond must also be specified, since different kinds of hydrogen bonds will generally have different energetic and geometric characteristics. In some cases, a fixed list of hydrogen bonds will be used throughout a simulation (a common example would be the Watson–Crick hydrogen bonds for base pairing in nucleic acids). In other cases, the hydrogen bonds are allowed to form and break spontaneously, in which case a provision is made to periodically examine the molecular configuration to see which potential donor–acceptor pairs satisfy appropriate criteria for hydrogen bonding. This is a standard procedure in simulations that include water.

The *energy parameter file* contains the parameters associated with each of the terms in the potential function. Each bond, for example, is characterized by a force constant k_b and an ideal length b_0. When the bond has a length b, the energy is

$$E_b = \frac{k_b}{2}(b-b_0)^2$$

The values of k_b and b_0 depend on the type of bond, so this file contains, for each bond type, the values of these parameters. Similarly, the parameters for the various types of bond angles, torsions, and hydrogen bonds must be given. For the nonbonded interactions, appropriate van der Waals parameters (A and C in equation 4.1) are also necessary, so this file will contain the value of those parameters for all pairs of atom types.

The *nonbonded pair list* is essential for the calculation of the nonbonded energy. This list is generated based on a rule for *nonbonded exclusions* plus a *cutoff distance* criterion. Normally, the interactions between pairs of atoms that are covalently bonded to one another (1–2 interactions) or that are separated by only two covalent bonds (1–3 interactions) are considered to be purely covalent in character, so 1–2 and 1–3 nonbonded interactions (van der Waals and electrostatic interactions) are excluded from the calculation of the molecular potential energy. Consequently, during the generation of the topology file, a list of these nonbonded exclusions is generated. Other nonbonded exclusion rules are common. For example, the nonbonded interactions between two atoms which are both part of an aromatic ring system (including atoms bonded directly to the ring) are normally not calculated.

The total number of terms in the potential energy function is dominated by the number of nonbonded interactions, so an enormous reduction in computer time can be obtained by ignoring nonbonded interactions for pairs of atoms that are separated by a distance greater than some reasonable cutoff distance. For a molecule with 1000 atoms, for example, there will be on the order of 1000 bonds, 1000 bond angles, and 1000 torsions, but there are about 500000 pairs of atoms. The combination of nonbonded exclusions and a cutoff radius of 0.5–1.0 nm will reduce the number of nonbonded pairs by about an order of magnitude. This choice of cutoff distance may be appropriate, because the electrostatic energy varies slowly with distance beyond that radius, and because the van der Waals interaction is very small beyond that distance (cf. section 4.2). Since the interatomic distances will change as the different parts of the molecule move relative to one another, this list must be updated periodically.

A2.2 Example: molecular dynamics simulation of phenylalanine tRNA

In this section we will describe the preparation of data files and the initiation procedure for carrying out a molecular dynamics simulation on a macromolecule, using $tRNA^{Phe}$ as an example. Because of space limitations, and because it is concepts rather than specific parameter values that we are trying to convey, we will not repeat the presentation of extensive tables of the parameters of the simulation. We will, instead, cite only a few typical values, referring the interested reader to the tables in the appendix of the original methods paper (Tung *et al.*, 1984) for details.

We begin by considering the information in the atom data file. There were thirteen distinct atom types used in the simulation, including phosphorus, three different types of oxygen atoms (sp^2 oxygen for carbonyl and nonester phosphate oxygens; sp^3 oxygen as in C–O–C and P–O–C structures; and an extended oxygen atom for hydroxyl groups), four different types of nitrogens, and five types of carbons. Table I of Tung *et al.* (1984) gives these atom types, plus the atomic polarizabilities, effective numbers of outer shell electrons, and atomic van der Waals radii, from which the various sets of nonbonded interaction parameters (A and C in equation (4.1)) were calculated for all pairs of atom types. These parameters were based on those given by Gelin (1976) and McCammon *et al.* (1979). The partial charges on each atom depend not only on atom type, but also on the local chemical structure. The values for these charges, Table V of Tung *et al.* (1984), were based on those of Miller (1979), with modification of the charges on the atoms of the phosphate group such that the net charge was -0.2 per residue, in units of the charge on the proton. Many protein simulations are done with a zero net charge on the molecule, i.e. at the isoelectric point, but nucleic acids bear one fully ionized phosphate group per residue at typical experimental pHs, producing a net charge which is only partially offset by counterion condensation at reasonable ionic strength. The effective net charge we chose is typical of those used for nucleic acid simulations and is close to that measured experimentally at physiological ionic strength (Olson *et al.*, 1976).

For the coordinate file, we used crystallographic coordinates from the Protein Data Bank, choosing those for $tRNA^{Phe}$ with the highest nominal resolution, 0.25 nm (Hingerty *et al.*, 1978).

Several steps are required to generate the topology file. The simplest procedure for macromolecules is to assemble a topology file for each type of residue (ribonucleotides for RNAs), then to string these together in the order dictated by the primary structure, and finally to perform any editing that is necessary. This final editing corrects the topology at the ends of the molecule (at the 5′ and 3′ ends of nucleic acids; at the amino and carboxy termini of proteins). It allows the specific modification of the topology file wherever special features occur (such as disulphide bridges in proteins). If explicit hydrogen bonds are to be included in the model, the list of these bonds is also established at this editing stage.

The residue topology files are originally generated manually, but once a library of these is generated – 20 amino acid files for protein studies and four deoxyribonucleotide files for DNA – they can be used to model any protein or DNA. In the case of $tRNA^{Phe}$, there are 15 distinct types of residues, because of chemical modifications (principally methylation) to both the bases and the riboses. The molecular topology file for $tRNA^{Phe}$ was generated by combining the individual residue topology files in the order specified by the primary structure.

Since this was one of the first molecular dynamics simulations on a nucleic acid, we wanted strong hydrogen bonds in order to stabilize the molecular structure. Consequently, a fixed list of explicit hydrogen bonds was used, and special hydrogen bonding parameters were introduced. For the hydrogen bond list, a trial set of 123 hydrogen bonds was generated automatically; it contained all donor–acceptor pairs within an arbitrary distance (0.325 nm). This list was then manually edited to contain only the 91 most commonly accepted crystallographic hydrogen bonds. The parameters for the hydrogen bonding potential function were determined by examining the shape of the total donor–acceptor potential energy curve for the sum of the van der Waals, electrostatic, and hydrogen bonding terms and requiring the minimum energy to have a value near -12 kJ/mol and to occur at a donor–acceptor distance near 0.3 nm. The set of 91 hydrogen bonds and the parameters of the hydrogen bonding interactions are given elsewhere (Harvey *et al.*, 1985).

With the exception of the hydrogen bonding parameters described in the previous paragraph, the $tRNA^{Phe}$ energy parameters (equation (4.1)) are typical of those used in most protein and nucleic acid modeling studies. They are given in the appendix of Tung *et al.* (1984).

Even when the starting structure for the simulation is that determined crystallographically, it is still necessary to relax the model using energy minimization prior to beginning the molecular dynamics simulation. This is because the crystallographic structure is itself a model, generated by balancing the desire for atomic positions that reproduce the observed diffraction pattern against the desire for a model with traditional values for bond lengths, bond angles, and so on. The initial round of energy minimization is necessary because the parameters for restraining internal coordinates during crystallographic refinement are generally not identical to those used in the molecular dynamics simulation, and because the crystal structure may contain nonbonded contact distances that are closer than those permitted by the parameters of the simulation. In the case of the $tRNA^{Phe}$ simulation, 100 cycles of steepest descent minimization reduced the potential energy of the molecule from over 13 000 kJ/mol to about 2500 kJ/mol, with the largest contribution coming from improved nonbonded geometries and a substantial contribution arising from improved planarities of the bases. After the equilibration procedure (described below), the total molecular potential energy had fallen to a value of about −5000 kJ/mol.

It should be noted that, in common with all potential energies, the molecular potential energy is measured relative to an arbitrary zero point, so the sign of the total energy does not have any significance. Only energy *differences* are relevant. Positive energies do not mean the structure is unstable. Conversely, a negative value for the potential energy does not necessarily indicate a stable structure.

After minimization, not all regions of the model are equally well refined, and this irregular distribution of potential energy must be removed during the warmup and equilibration period. The $tRNA^{Phe}$ model was heated from 50 K to 300 K over a period of 11 ps, followed by 5 ps during which random thermal velocities corresponding to a Maxwellian distribution with a mean temperature of 300 K were repeatedly reassigned at intervals of 0.2 ps. Further equilibration took place over a period of 4 ps, during which the simulation was run without further reassignment of velocities. It is customary to begin data analysis only after such a free run, in order to be certain that an equilibrium situation has been adequately approximated; this is done by verifying that there are no systematic drifts of potential energy and that there are no serious changes in the large-scale structure of the model during the free run.

APPENDIX 3

Molecular dynamics at constant temperature and pressure (Andersen, 1980; Berendsen *et al.* 1984)

Molecular dynamics has an important advantage over Monte Carlo methods, in that it provides information about the time-dependence and magnitude of fluctuations in both positions and velocities, while Monte Carlo provides only positional information and gives no information on time-dependence. The original formulations of molecular dynamics had two shortcomings, however. First, simulations on fluids or solutions were carried out at constant volume, although volume fluctuations are critical to a number of phenomena. Second, the numerical integration of the equations of motion produces a trajectory in which the total energy of the system is conserved, while real systems generally exchange energy with their surroundings. A normal molecular dynamics free run corresponds to a simulation on the microcanonical ensemble, and it is often necessary to simulate the behavior of other ensembles. Methods for simulating other ensembles using Monte Carlo had been available for several years (Wood, 1968; Valleau & Whittington, 1977; Valleau & Torrie, 1977), and it was Andersen (1980) who first devised methods for molecular dynamics simulations on the other ensembles listed in table A3.1.

There are a number of situations where free molecular dynamics may be inappropriate and the methods described below are useful. For example, experiments on fluids and solutions are usually made at either fixed temperature and pressure or at fixed temperature and density, and appropriate ensembles should be used for simulations on these systems. Spontaneous fluctuations in both energy and volume must be allowed if one is looking at phenomena like nucleation or phase separation. If there are reactions that take up or release energy, exchange of heat with the surroundings is necessary if large temperature changes are to be avoided within the volume being simulated. Finally, there may be cases where it

Table A3.1. *Molecular dynamics simulations of various ensembles*

Condition	Constants during simulation*	Ensemble name
Free run	N, E, V	Microcanonical
Constant temperature	N, T, V	Canonical
Constant pressure	N, P, H	Isobaric–Isoenthalpic
Constant temperature and pressure	N, T, P	Isothermal–Isobaric

* N = number of particles; E = energy; T = temperature; P = pressure; V = volume; H = enthalpy

is desired to heat or cool the system, or to compress or expand it; these simulations would allow the determination of properties like the isobaric heat capacity or the compressibility (either adiabatic or isothermal).

This appendix describes the modifications to the normal free molecular dynamics algorithm that allow one to simulate these other ensembles. The full theoretical justification for these procedures is given in the original paper by Andersen (1980), and only the procedures themselves are described here.

A3.1 Molecular dynamics at constant temperature

A small volume of matter in thermal equilibrium with a surrounding constant temperature bath exchanges energy with the bath. To simulate this exchange of energy, stochastic impulses are given to each particle in the system being treated. Each impulse instantaneously changes the momentum of the particle; the frequency of these collisions, and the resulting momentum probability distribution function, must be chosen such that the average value of any property, calculated over a very long trajectory, will be the equal to the canonical ensemble average of that same property.

To carry out the simulation, values are chosen for T, the desired temperature, and ν, the mean frequency with which each particle experiences a stochastic collision. The probability that any particular particle suffers a collision in the time interval Δt is $\nu\Delta t$. The times at which each particle is to suffer a collision can be determined using random numbers to calculate the intervals between successive collisions, noting that this is a Poisson process, so that the probability distribution function for the length of the interval, t, is distributed exponentially.

$$P(t) = \nu\, e^{-\nu t}$$

The simulation is started by assigning initial positions and velocities as in a normal free molecular dynamics run, and the equations of motion are integrated until the time of the first stochastic collision. The particle suffering the collision is given a new velocity, chosen at random from the Boltzmann distribution at temperature T. This instantaneous change of momentum does not affect the instantaneous positions or velocities of the other particles in the system. The equations of motion are integrated until the time for the next collision, and this process of short free runs interrupted by instantaneous velocity reassignments is repeated throughout the duration of the simulation.

How should ν, the frequency of stochastic collisions, be chosen? Andersen (1980) has shown that ν is inversely proportional to $N^{\frac{2}{3}}$, where N is the number of particles in the system, according to the relationship

$$\nu = \nu_c / N^{\frac{2}{3}} \tag{A3.1}$$

where ν_c would be the collision frequency if the sample contained only one particle. For macromolecular systems, the particles are atoms (or extended atoms), and the collision frequency for any particular atom is on the order of the frequency of covalent bond vibrations, i.e. $\nu_c \approx 10^{14}\ s^{-1}$. For a typical macromolecule with a molecular weight of about 10^4, N is of the order of 1000 heavy atoms, and ν is about 1 ps. This is an attractive coincidence, because in many molecular dynamics simulations, the time step Δt is on the order of 0.001 ps. Consequently, at each time step one atom is chosen at random for a stochastic collision. Over a period of 1 ps (1000 time steps), each atom will suffer an average of one collision, as required. This same procedure can generally be used on systems containing from 100 to 10000 atoms without serious error, because the large number of atoms guarantees that differences between ensemble averages will be small (McQuarrie, 1976).

An alternative method, due to Berendsen *et al.* (1984), couples the system to a bath whose temperature is T_0 by rescaling all atomic velocities at each time step. The velocity v is scaled to a value λv, where

$$\lambda = 1 + \frac{\Delta t}{2\tau_T}(T_0/T - 1)$$

In this expression, Δt is the size of the time step, τ_T is a characteristic relaxation time, and T is the instantaneous temperature. If one prefers to require the change in temperature per time step to be exactly equal to $(T_0 - T)\Delta t/\tau_T$, one can use

$$\lambda = \left[1 + \frac{\Delta t}{\tau_T}\left(\frac{T_0}{T} - 1\right)\right]^{\frac{1}{2}}$$

The advantage of this procedure over that of Anderson (1980) is that the strength of coupling can be chosen (by choosing the value of τ_T) to suit the needs of a given application. On the other hand, the simulation does not correspond exactly to any of the standard thermodynamic ensembles.

A3.2 Molecular dynamics at constant pressure

In a normal molecular dynamics simulation with repeating boundary conditions, the volume is held fixed, whereas at constant pressure the volume of the system fluctuates. To include the volume as a dynamic variable, the equations of motion are determined in an analysis of a system in which the positions and momenta of all the particles are scaled by a factor proportional to the cube root of the volume of the system (Anderson, 1980). If we denote the volume V, the resulting equations, which replace equations (4.15) and (4.16), are, for each real atom i,

$$\dot{\mathbf{r}}_i = \mathbf{p}_i/m_i + V^*\mathbf{r}_i/3$$

$$\dot{\mathbf{p}}_i = \Sigma_j \, \mathbf{F}_{ij} - V^*\mathbf{p}_i/3$$

$$M\ddot{V} = -\alpha + [\Sigma_i \, (\mathbf{p}_i \cdot \mathbf{p}_i)/m_i - \Sigma_{ij} \, \mathbf{r}_{ij} \, U'(\mathbf{r}_{ij})]/(3V)$$

where U' represents the derivative of the potential with respect to the position of atom i, and

$$V^* = \frac{\mathrm{d}}{\mathrm{d}t}(\ln V)$$

In these equations, the sums are to be calculated as follows. In the second equation, the summation is over all values of $j \neq i$, and the minimum image convention is followed: for a given value of j, the pairwise ij interaction is only considered between atom i and whichever is closest to it, the real atom j or the closest image of atom j. Both sums in the third equation are taken only over real atoms, and the second sum is only calculated for values of $j > i$. This is to guarantee that volume fluctuations reflect the imbalance due to the external pressure α and the pressure arising from interatomic interactions within the volume. At each step of the dynamics, distances for repeating boundary conditions are rescaled so that the sample pressure has the correct value.

There are two parameters that must be determined at the outset of the simulation, α and M. The former is simply the pressure at which the simulation is to be carried out, while M can be interpreted as the effective mass of a piston that applies the pressure α. The trajectory averages calculated from the simulation will be independent of M as long as M is finite and positive, and any value of M can be used if all that is to be determined is a set of equilibrium averages. If, however, an accurate representation of the system fluctuations is desired, it is important to choose a physically appropriate value for M. While no precise formulation is available for the specification of M (Haile & Graben, 1980), a value can be determined by trial and error by following the prescription that M should be chosen such that the time scale for the fluctuations in the volume is roughly equal to $V^{\frac{1}{3}}$ divided by the speed of sound in the fluid (Andersen, 1980).

To verify that the numerical integration of the equations of motion is being correctly carried out in this system, it should be noted that the quantity that is conserved is the Hamiltonian composed of terms for the kinetic and potential energy of the particles within the volume, plus the kinetic and potential energy of the piston. This Hamiltonian is

$$H = \tfrac{1}{2}\Sigma_i\,(\mathbf{p}_i \cdot \mathbf{p}_i/m_i) + \Sigma_{ij}\, U(r_{ij}) + \tfrac{1}{2}M\dot{V}^2 + \alpha V$$

In the expression, both sums are calculated only over real particles; the second sum is again calculated only over values of $j > i$, so it is equal to the total interatomic potential energy.

Under some circumstances, the previous algorithm has a tendency to produce oscillatory behavior in response to pressure changes. An alternative procedure rescales the coordinates of each atom at each time step (Berendsen *et al.*, 1984). The atomic coordinate, x, and the characteristic distance for repeating boundary conditions, d, are rescaled to values μx and μd, respectively, where

$$\mu = \left[1 - \frac{\Delta t}{\tau_P}(P_0 - P)\right]^{\frac{1}{3}}$$

Here, Δt is the size of the time step, τ_P is a characteristic relaxation time, and P_0 is the pressure of the external constant pressure bath. The instantaneous pressure can be calculated as follows:

$$P = \frac{2}{3V}\left[E_k + \frac{1}{2}\Sigma_{i<j}\,\mathbf{r}_{ij} \cdot \mathbf{F}_{ij}\right]$$

where V is the volume and E_k is the kinetic energy, $\mathbf{r}_{ij}$ is the vector from particle i to particle j, and $\mathbf{F}_{ij}$ is the force on particle j due to particle i. The details of the algorithm for simultaneously coupling pressure and temperature to an external bath are given by Berendsen *et al.* (1984). Additional references are available in Evans & Holian (1985).

REFERENCES

Abola, E. E., Bernstein, F. C. & Koetzle, T. F. (1985). 'The protein data bank.' In *Molecular Dynamics and Protein Structure*, ed. J. Hermans, pp. 167–9. Chapel Hill: University of North Carolina.

Adam, G. & Delbrück, M. (1968). 'Reduction of dimensionality in biological diffusion processes.' In *Structural Chemistry and Molecular Biology*, eds. A. Rich & N. Davidson, p. 198. San Francisco: Freeman.

Adams, D. J. (1979). 'Computer simulation of ionic systems: The distorting effects of the boundary conditions.' *Chemical Physics Letters*, **62**, 329.

Agmon, N. & Hopfield, J. J. (1983). 'CO binding to heme proteins: A model for barrier height distributions and slow conformational changes.' *Journal of Chemical Physics*, **79**, 2042.

Alagona, G., Ghio, C. & Kollman, P. A. (1985). 'Monte Carlo simulations of the solvation of the dimethyl phosphate anion.' *Journal of the American Chemical Society*, **107**, 2229.

Alder, B. J. & Wainwright, T. E. (1960). 'Studies in molecular dynamics II. Behavior of a small number of elastic spheres.' *Journal of Chemical Physics*, **33**, 1439.

Aleksandrov, I. V. & Goldanskii, V. I. (1984). 'On the dependence of the rate constant of an elementary chemical process in the kinetic region on the viscosity of the medium.' *Chemical Physics*, **87**, 455.

Allison, S. A., Ganti, G. & McCammon, J. A. (1985*a*). 'Simulation of the diffusion-controlled reaction between superoxide and superoxide dismutase. I. Simple models.' *Biopolymers*, **24**, 1323.

Allison, S. A. & McCammon, J. A. (1984*a*). 'Transport properties of rigid and flexible macromolecules by Brownian dynamics simulation.' *Biopolymers*, **23**, 167.

Allison, S. A. & McCammon, J. A. (1984*b*) 'Multistep Brownian dynamics: Application to short wormlike chains.' *Biopolymers*, **23**, 363.

Allison, S. A. & McCammon, J. A. (1985). 'Dynamics of substrate binding to copper zinc superoxide dismutase.' *Journal of Physical Chemistry*, **89**, 1072.

Allison, S. A., Northrup, S. H. & McCammon, J. A. (1985*b*). 'Extended Brownian dynamics of diffusion controlled reactions.' *Journal of Chemical Physics*, **83**, 2894.

Allison, S. A., Northrup, S. H. & McCammon, J. A. (1986). 'Simulation of biomolecular diffusion and complex formation.' *Biophysical Journal*, **49**, 167.

Allison, S. A., Srinivasan, N., McCammon, J. A. & Northrup, S. H. (1984). 'Diffusion-controlled reactions between a spherical target and dumbell dimer by Brownian dynamics simulation.' *Journal of Physical Chemistry*, **88**, 6152.

Altman, S., ed. (1978). *Transfer RNA*. Cambridge, Massachusetts: The MIT Press.

Altona, C. & Sundaralingam, M. (1972). 'Conformational analysis of the sugar ring in nucleosides and nucleotides. A new description using the concept of pseudorotation.' *Journal of the American Chemical Society*, **94**, 8205.

Andersen, H. C. (1980). 'Molecular dynamics simulations at constant pressure and/or temperature.' *Journal of Chemical Physics*, **72**, 2384.

Anderson, C. M., Zucker, F. H. & Steitz, T. A. (1979). 'Space-filling models of kinase clefts and conformation changes.' *Science*, **204**, 375.

Anderson, J. E., Ptashne, M. & Harrison, S. C. (1985). A phage repressor-operator complex at 7 Å resolution, *Nature*, **316**, 596.

Anfinsen, C. B. (1973). 'Principles that govern the folding of protein chains.' *Science*, **181**, 223.

Ansari, A., Berendzen, J., Browne, S. F., Frauenfelder, H., Iben, I. E. T., Sauke, T. B., Shyamsunder, E. & Young, R. D. (1985). 'Protein states and proteinquakes.' *Proceedings of the National Academy of Sciences, USA*, **82**, 5000.

Åqvist, J., van Gunsteren, W. F., Leijonmarck, M. & Tapia, O. (1985). 'A molecular dynamics study of the C-terminal fragment of the L7/L12 ribosomal protein: Secondary structure motion in a 150 ps trajectory.' *Journal of Molecular Biology*, **183**, 461.

Arnott, S., Campbell Smith, P. J. & Chandrasekaran, R. (1976). 'Atomic coordinates and molecular conformations for DNA–DNA, RNA–RNA, and DNA-RNA helices.' *Handbook of Biochemistry and Molecular Biology*, ed. G. D. Fasman, pp. 411–22. Cleveland, Ohio: Chemical Rubber Company.

Artymiuk, P. J., Blake, C. C. F., Grace, D. E. P., Oatley, S. J., Phillips, D. C. & Sternberg, M. J. E. (1979). 'Crystallographic studies of the dynamic properties of lysozyme.' *Nature*, **280**, 563.

Austin, R. H., Beeson, K. W., Eisenstein, L., Frauenfelder, H. & Gunsalus, I. C. (1975). 'Dynamics of ligand binding to myoglobin.' *Biochemistry*, **14**, 5355.

Bacquet, R. J. & McCammon, J. A. (1986). 'Salt effects in enzyme-substrate interactions: Monte Carlo simulation.' *Annals of the New York Academy of Sciences*, in press.

Bacquet, R. & Rossky, P. J. (1984). Ionic atmosphere of rodlike polyelectrolytes. A hypernetted chain study.' *Journal of Physical Chemistry*, **88**, 2660.

Baker, E. N. & Hubbard, R. E. (1984). 'Hydrogen bonding in globular proteins.' *Progress in Biophysics and Molecular Biology*, **44**, 97.

Banaszak, L. J., Birktoft, J. J. & Barry, C. D. (1981). 'Protein–protein interactions and protein structures.' In *Protein–Protein Interactions*, eds. C. Frieden & L. W. Nichol, p. 31. New York: Wiley.

Banner, D. W., Bloomer, A. C., Petsko, G. A., Phillips, D. C., Pogson, C. I.,

Wilson, I. A., Corran, P. H., Furth, A. J., Milman, J. D., Offord, R. E., Priddle, J. D. & Waley, S. G. (1975). 'Structure of chicken muscle triose phosphate isomerase determined crystallographically at 2.5 Å resolution using amino acid sequence data.' *Nature*, **255**, 609.

Barkley, M. D., & Zimm, B. H. (1979). 'Theory of twisting and bending of chain macromolecules; analysis of the fluorescence depolarization of DNA.' *Journal of Chemical Physics*, **70**, 2991.

Barry, D. & Crowther, W. (1985). 'The butterfly microprocessor.' In *Molecular Dynamics and Protein Structure*, ed. J. Hermans, pp. 191–3. Chapel Hill: University of North Carolina.

Bartunik, H. D., Jollès, P., Berthou, J. & Dianoux, A. J. (1982). 'Intramolecular low-frequency vibrations in lysozyme by neutron time-of-flight spectroscopy.' *Biopolymers*, **21**, 43.

Bauminger, E. R., Cohen, S. G., Nowik, I., Ofer, S. & Yariv, J. (1983). 'Dynamics of heme iron in crystals of metmyoglobin and deoxymyoglobin.' *Proceedings of the National Academy of Sciences of the USA*, **80**, 736.

Becker, K. D. & Hoheisel, C. (1982). 'The dynamic behavior of a crystal during defect jumps. I. Molecular dynamics studies using Lennard–Jones potential functions.' *Journal of Chemical Physics*, **77**, 5108.

Beddell, C. R. (1984). 'Designing drugs to fit a macromolecular receptor.' *Chemical Society Reviews*, **13**, 279.

Beece, D., Eisenstein, L., Frauenfelder, H., Good, D., Marden, M. C., Reinisch, L., Reynolds, A. H., Sorensen, L. B. & Yue, K. T. (1980). 'Solvent viscosity and protein dynamics.' *Biochemistry*, **19**, 5147.

Beechem, J. M. & Brand, L. (1985). 'Time-resolved fluorescence of proteins.' *Annual Review of Biochemistry*, **54**, 43.

Beeman, D. (1976). 'Some multistep methods for use in molecular dynamics calculations.' *Journal of Computational Physics*, **20**, 130.

Belch, A. C. & Berkowitz, M. (1985). 'Molecular dynamics simulations of TIPS2 water restricted by a spherical hydrophobic boundary.' *Chemical Physics Letters*, **113**, 278.

Bell, C. G. (1985). 'Multis: A new class of multiprocessor computers.' *Science*, **228**, 462.

Ben-Naim, A. (1974). *Water and Aqueous Solutions*. New York: Plenum.

Ben-Naim, A. (1980). *Hydrophobic Interactions*. New York: Plenum.

Bennett, C. H. (1975). 'Exact defect calculations in model substances.' In *Diffusion in Solids*, ed, J. J. Burton & A. S. Nowick, pp. 73–113. San Francisco: Academic.

Bennett, C. H. (1976). 'Efficient estimation of free energy differences from Monte Carlo data.' *Journal of Computational Physics*, **22**, 245.

Bennett, W. S. & Huber, R. (1984). 'Structural and functional aspects of domain motions in proteins.' *CRC Critical Reviews in Biochemistry*, **15**, 291.

Berendsen, H. J. C. (1985*a*). 'Treatment of long-range forces in molecular dynamics.' In *Molecular Dynamics and Protein Structure*, ed. J. Hermans, pp. 18–22. Chapel Hill: University of North Carolina.

Berendsen, H. J. C. (1985*b*). 'Statistical mechanics and molecular dynamics: The calculation of free energy.' In *Molecular Dynamics and Protein Structure*, ed. J. Hermans, pp. 43–6. Chapel Hill: University of North Carolina.

Berendsen, H. J. C. (1985*c*). 'Molecular dynamics on vector processors and special purpose computers.' In *Molecular Dynamics and Protein Structure*, ed. J. Hermans, pp. 123–5. Chapel Hill: University of North Carolina.

Berendsen, H. J. C., Postma, J. P. M., van Gunsteren, W. F., DiNola, A. & Haak, J. R. (1984). 'Molecular dynamics with coupling to an external bath.' *Journal of Chemical Physics*, **81**, 3684.

Berendsen, H. J. C., Postma, J. P. M., van Gunsteren, W. F. & Hermans, J. (1981). 'Interaction models for water in relation to protein hydration.' In *Intermolecular Forces*, ed. B. Pullman, pp. 331–42. Holland: Reidel.

Berendsen, H. J. C., van Gunsteren, W. F. & Postma, J. P. M. (1984). 'Molecular dynamics on Cray, CYBER and DAP.' In *High-Speed Computation*, ed. J. S. Kowalik, p. 425. Berlin: Springer-Verlag.

Berendsen, H. J. C., van Gunsteren, W. F., Zwinderman, H. R. J. & Geurtsen, R. G. (1986). 'Simulations of proteins in water.' *Annals of the New York Academy of Sciences*, in press.

Berens, P. H., Mackay, D. H. J., White, G. M. & Wilson, K. R. (1983). 'Thermodynamics and quantum corrections from molecular dynamics for liquid water.' *Journal of Chemical Physics*, **79**, 2375.

Berg, O. G. (1984). 'Diffusion-controlled protein-DNA association: Influence of segmental diffusion of the DNA.' *Biopolymers*, **23**, 1869.

Berg, O. G. & von Hippel, P. H. (1985). 'Diffusion-controlled macromolecular interactions.' *Annual Review of Biophysics and Biophysical Chemistry*, **14**, 131.

Berg, O. G., Winter, R. B. & von Hippel, P. H. (1981). 'Diffusion-driven mechanisms of protein translocation on nucleic acids. 1. Models and theory.' *Biochemistry*, **20**, 6929.

Berkowitz, M., Karim, O. A., McCammon, J. A. & Rossky, P. J. (1984). 'Sodium chloride ion pair interaction in water: Computer simulation.' *Chemical Physics Letters*, **105**, 577.

Berkowitz, M. & McCammon, J. A. (1981). 'Brownian motion of a system of coupled harmonic oscillators.' *Journal of Chemical Physics*, **75**, 957.

Berkowitz, M. & McCammon, J. A. (1982). 'Molecular dynamics with stochastic boundary conditions.' *Chemical Physics Letters*, **90**, 215.

Berkowitz, M., Morgan, J. D., McCammon, J. A. & Northrup, S. H. (1983). 'Diffusion-controlled reactions: A variational formula for the optimum reaction coordinate.' *Journal of Chemical Physics*, **79**, 5563.

Bernstein, F. C., Koetzle, T. F., Williams, G. J. B., Meyer, E. F., Brice, M. D., Rodgers, J. R., Kennard, O., Shimanouchi, T. & Tasumi, M. (1977). 'The protein data bank: A computer-based archival file for macromolecular structures.' *Journal of Molecular Biology*, **112**, 535.

Berry, R. S., Rice, S. A., & Ross, J. (1980). *Physical Chemistry*. New York: John Wiley.

Berzofsky, J. A. (1985). 'Intrinsic and extrinsic factors in protein antigenic structure.' *Science*, **229**, 932.

Beveridge, D. L., Maye, P. V., Jayaram, B., Ravishanker, G. & Mezei, M. (1984). 'Aqueous hydration of nucleic acid constituents: Monte Carlo computer simulation studies.' *Journal of Biomolecular Structure and Dynamics*, **2**, 261.

Beveridge, D. & Mezei, M. (1985). 'Free energy simulations: The coupling parameter approach and topographical transition coordinates.' In *Molecular Dynamics and Protein Structure*, ed. J. Hermans, pp. 53–7. Chapel Hill: University of North Carolina.

Beveridge, D. L., Mezei, M., Mehrotra, P. K., Marchese, F. T., Ravishanker, G., Vasu, T. & Swaminathan, S. (1983). 'Monte Carlo computer simulation studies of the equilibrium properties and structure of liquid water.' In *Molecular-Based Study of Fluids*, eds. J. M. Haile & G. A. Mansoori, pp. 297–351. Washington, D.C.: American Chemical Society.

Bloomfield, V., Dalton, W. O. & van Holde, K. E. (1967). 'Frictional coefficients of multisubunit structures.' I. Theory. *Biopolymers*, **5**, 135.

Bockris, J. O'M. & Reddy, A. K. N. (1970). *Modern Electrochemistry*. New York: Plenum.

Bode, W. (1979). 'The transition of bovine trypsinogen to a trypsin-like state upon strong ligand binding. II. The binding of the pancreatic trypsin inhibitor and of isoleucine-valine and of sequentially related peptides to trypsinogen and to p-guanidinobenzoate-trypsinogen.' *Journal of Molecular Biology*, **127**, 357.

Bounds, D. G. (1985). 'A molecular dynamics study of the structure of water around the ions Li^+, Na^+, K^+, Ca^{++}, Ni^{++} and Cl^-.' *Molecular Physics*, **54**, 1335.

Brant, D. A. & Flory, P. J. (1965). 'The configuration of random polypeptide chains.' II. Theory. *Journal of the American Chemical Society*, **87**, 2791.

Brant, D. A., Miller, W. G. & Flory, P. J. (1967). 'Conformational energy estimates for statistically coiling polypeptide chains.' *Journal of Molecular Biology*, **23**, 47.

Brooks, B. & Karplus, M. (1983). 'Harmonic dynamics of proteins: Normal modes and fluctuations in bovine pancreatic trypsin inhibitor.' *Proceedings of the National Academy of Sciences of the USA*, **80**, 6571.

Brooks, B. & Karplus, M. (1985). 'Normal modes for specific motions of macromolecules: Application to the hinge-bending mode of lysozyme.' *Proceedings of the National Academy of Sciences of the USA*, **82**, 4995.

Brooks, B. R., Bruccoleri, R. E., Olafson, B. D., States, D. J., Swaminathan, S. & Karplus, M. (1983). 'CHARMM: A program for macromolecular energy, minimization, and dynamics calculations.' *Journal of Computational Chemistry*, **4**, 187.

Brooks, C. L., Brünger, A. & Karplus, M. (1985). 'Active site dynamics in protein molecules: A stochastic boundary molecular-dynamics approach.' *Biopolymers*, **24**, 843.

Brooks, C. L., Pettitt, B. M. & Karplus, M. (1985). 'Structural and energetic effects of truncating long ranged interactions in ionic and polar fluids.' *Journal of Chemical Physics*, **83**, 5897.

Brown, K. G., Erfurth, S. C., Small, E. W. & Peticolas, W. L. (1972). 'Conformationally dependent low-frequency motions of proteins by laser Raman spectroscopy.' *Proceedings of the National Academy of Sciences of the USA*, **69**, 1467.

Bruccoleri, R. E., Karplus, M. & McCammon, J. A. (1986). 'The hinge bending mode of a lysozyme–inhibitor complex.' *Biopolymers*, **25**, 1767.

Brünger, A. T., Brooks, C. L. & Karplus, M. (1985). 'Active site dynamics of ribonuclease.' *Proceedings of the National Academy of Sciences of the USA*, **82**, 8458.

Burkert, U. & Allinger, N. L. (1982). *Molecular Mechanics*. Washington, D.C.: American Chemical Society.

Butz, T., Lerf, A. & Huber, R. (1982). 'Intramolecular reorientational motion in trypsinogen studied by perturbed angular correlation of ^{199m}Hg labels.' *Physical Review Letters*, **48**, 890.

Calef, D. F. & Deutch, J. M. (1983). 'Diffusion-controlled reactions.' *Annual Review of Physical Chemistry*, **34**, 493.

Calef, D. F. & Wolynes, P. G. (1983). 'Classical solvent dynamics and electron transfer. II. Molecular aspects.' *Journal of Chemical Physics*, **78**, 470.

Calhoun, D. B., Vanderkooi, J. M. & Englander, S. W. (1983*a*). 'Penetration of small molecules into proteins studied by quenching of phosphorescence and fluorescence.' *Biochemistry*, **22**, 1533.

Calhoun, D. B., Vanderkooi, J. M., Woodrow, G. V. & Englander, S. W. (1983*b*). 'Penetration of dioxygen into proteins studied by quenching of phosphorescence and fluorescence.' *Biochemistry*, **22**, 1526.

Calladine, C. R. (1982). 'Mechanics of sequence-dependent stacking of bases in B-DNA.' *Journal of Molecular Biology*, **161**, 343.

Calladine, C. R. & Drew, H. R. (1984). 'A base-centred explanation of the B-to-A transition in DNA.' *Journal of Molecular Biology*, **178**, 773.

Cantor, C. R. & Schimmel, P. R. (1980). *Biophysical Chemistry*. San Francisco: Freeman.

Careri, G., Fasella, P. & Gratton, E. (1979). 'Enzyme dynamics: The statistical physics approach.' *Annual Review of Biophysics and Bioengineering*, **8**, 69.

Carotti, A., Hansch, C., Mueller, M. M. & Blaney, J. M. (1984). 'Actinidin hydrolysis of substituted-phenyl hippurates: A quantitative structure-activity relationship and graphics comparison with hydrolysis by papain.' *Journal of Medicinal Chemistry*, **27**, 1401.

Carson, M. & Hermans, J. (1985). 'The molecular dynamics workshop laboratory.' In *Molecular Dynamics and Protein Structure*, ed. J. Hermans, pp. 165–6. Chapel Hill: University of North Carolina.

Case, D. A. & Karplus, M. (1979). 'Dynamics of ligand binding to heme proteins.' *Journal of Molecular Biology*, **132**, 343.

Case, D. A. & McCammon, J. A. (1986). 'Dynamical simulations of oxygen binding to myoglobin.' *Annals of the New York Academy of Sciences*, in press.

Cavailles, J. A., Neumann, J. M., Taboury, J., d'Estaintot, B. L., Huynh-Dinh, T., Igolen, J. & Tran-Dinh, S. (1984). 'B, Z conformations and mechanism of the Z–B–coil transitions of the self-complementary deoxy-hexanucleotide d(C-G-m^{5}C-G-C-G) by ^{1}H-NMR and CD spectroscopy.' *Journal of Biomolecular Structure and Dynamics*, **1**, 1347.

Chaires, J. B., Dattagupta, N. & Crothers, D. M. (1985). 'Kinetics of the daunomycin–DNA interaction.' *Biochemistry*, **24**, 260.

Chakerian, A. E., Pfahl, M., Olson, J. S. & Matthews, K. S. (1985). 'A mutant lactose repressor with altered inducer and operator binding parameters'. *Journal of Molecular Biology*, **183**, 43.

Chandler, D. (1978). 'Statistical mechanics of isomerization dynamics in liquids and the transition state approximation.' *Journal of Chemical Physics*, **68**, 2959.

Chandler, D. (1982). 'Equilibrium theory of polyatomic fluids.' In *The Liquid State of Matter: Fluids, Simple and Complex*, eds E. W. Montroll & J. L. Lebowitz, pp. 275–340. Amsterdam: North-Holland.

Chandler, D. (1984). 'Quantum theory of solvation.' *Journal of Physical Chemistry*, **88**, 3400.

Chandrasekhar, J., Smith, S. F. & Jorgensen, W. L. (1985). 'Theoretical examination of the S_N2 reaction involving chloride ion and methyl chloride in the gas phase and aqueous solution.' *Journal of the American Chemical Society*, **107**, 154.

Chandrasekhar, S. (1943). 'Stochastic problems in physics and astronomy.' *Reviews of Modern Physics*, **15**, 1.

Chiles, R. A. & Rossky, P. J. (1984). 'Evaluation of reaction free energy surfaces in aqueous solution: An integral equation approach.' *Journal of the American Chemical Society*, **106**, 6867.

Chothia, C. (1984). 'Principles that determine the structure of proteins.' *Annual Review of Biochemistry*, **53**, 537.

Chou, K. & Zhou, G. (1982). 'Role of protein outside active site on the diffusion-controlled reaction of enzyme.' *Journal of the American Chemical Society*, **104**, 1409.

Clementi, E. & Corongiu, G. (1981). 'Simulations of the solvent structure for macromolecules: Solvation model for B-DNA and Na^+-B-DNA double helix at 300° K.' *Annals of the New York Academy of Sciences*, **367**, 83.

Clementi, E., Corongiu, G., Jönsson, B. & Romano, S. (1979). 'The water structure in the active cleft of human carbonic anhydrase B.' *Federation of European Biochemical Societies Letters*, **100**, 313.

Clementi, E. & Sarma, R. H. eds. (1983). *Structure and Dynamics: Nucleic Acids and Proteins*. New York: Adenine Press.

Clore, G. M., Gronenborn, A. M., Brünger, A. T. & Karplus, M. (1985). 'Solution conformation of a heptadecapeptide comprising the DNA binding helix F of the cyclic AMP receptor protein of Escherichia coli. Combined use of ^1H nuclear magnetic resonance and restrained molecular dynamics.' *Journal of Molecular Biology*, **186**, 435.

Connolly, M. L. (1985). 'Computation of molecular volume.' *Journal of the American Chemical Society*, **107**, 1118.

Conway, B. E. (1981). *Ionic Hydration in Chemistry and Biophysics*. Oxford: Elsevier.

Cooper, A. (1984). 'Protein fluctuations and the thermodynamic uncertainty principle.' *Progress in Biophysics and Molecular Biology*, **44**, 181.

Craik, C. S., Largman, C., Fletcher, T., Roczniak, S., Barr, P. J., Fletterick, R., Rutter, W. J. (1985). 'Redesigning trypsin: Alteration of substrate specificity.' *Science*, **228**, 291.

Creighton, T. E. (1983). *Proteins*. New York: Freeman.

Creighton, T. E. (1985). 'The problem of how and why proteins adopt folded conformations.' *Journal of Physical Chemistry*, **89**, 2452.

Cremer, D. & Pople, J. A. (1975). 'A general definition of ring puckering coordinates.' *Journal of the American Chemical Society*, **97**, 1354.

Crick, F. H. C. (1966). 'Codon–anticodon pairing: The wobble hypothesis.' *Journal of Molecular Biology*, **19**, 548.

Crothers, D. M. (1979). 'Physical studies of tRNA in solution.' In *Transfer RNA: Structure, Properties, and Recognition*, eds P. R. Schimmel, D. Söll & J. N. Abelson. New York: Cold Spring Harbor.

Cudd, A. & Fridovich, I. (1982). 'Electrostatic interactions in the reaction mechanism of bovine erythrocyte superoxide dismutase.' *Journal of Biological Chemistry*, **257**, 11443.

Daune, M. P., Westhof, E., Koffel-Schwartz, N. & Fuchs, R. P. P. (1985). 'Covalent binding of a carcinogen as a probe for the dynamics of deoxyribonucleic acid.' *Biochemistry*, **24**, 2275.

Debrunner, P. G. & Frauenfelder, H. (1982). 'Dynamics of proteins.' *Annual Review of Physical Chemistry*, **33**, 283.

de Gennes, P. G. & Papoular, M. (1969). *Polarization, Matière et Rayonnement*. Paris: Presses Universitaire de France.

Depew, R. E. & Wang, J. C. (1975). 'Conformational fluctuations of DNA helix.' *Proceedings of the National Academy of Sciences of the USA*, **72**, 4275.

Dickerson, R. E. (1983). 'Base sequence and helix structure variation in B and A DNA.' *Journal of Molecular Biology*, **166**, 419.

Dickerson, R. E., Drew, H. R., Conner, B. N., Wing, R. M., Fratini, A. V. & Kopka, M. L. (1982). 'The anatomy of A-, B- and Z-DNA.' *Science*, **216**, 475.

Dickinson, E. (1985). 'Brownian dynamics with hydrodynamic interactions: The application to protein diffusional problems.' *Chemical Society Reviews*, **14**, 421.

Dickinson, E., Allison, S. A. & McCammon, J. A. (1985). 'Brownian dynamics with rotation–translation coupling.' *Journal of the Chemical Society, Faraday Transactions* II, **81**, 591.

Dickinson, E. & Honary, F. (1986). 'A Brownian dynamics simulation of enzyme–substrate encounters at the surface of a colloidal particle.' *Journal of the Chemical Society, Faraday Transactions* II, in press.

Di Nola, A., Berendsen, H. J. C. & Edholm, O. (1984). 'Free energy determination of polypeptide conformations generated by molecular dynamics.' *Macromolecules*, **17**, 2044.

Dixon, L. C. W. (1972). *Nonlinear Optimisation*. New York: Crane, Russak.

Drew, H. R., Samson, S. & Dickerson, R. E. (1982). 'Structure of a B-DNA dodecamer at 16 K.' *Proceedings of the National Academy of Sciences of the USA*, **79**, 4040.

Drew, H. R., Wing, R. M., Takano, T., Broka, C., Tanaka, S., Itakura, K. & Dickerson, R. E. (1981). 'Structure of a B-DNA dodecamer: Conformation and dynamics.' *Proceedings of the National Academy of Sciences of the USA*, **78**, 2179.

Edholm, O. & Berendsen, H. J. C. (1984). 'Energy estimation from simulations of non-diffusive systems.' *Molecular Physics*, **51**, 1011.

Edholm, O., Nilsson, L., Berg, O., Ehrenberg, M., Claesens, F., Gräslund, A., Jönsson, B. & Teleman, O. (1984). 'Biomolecular dynamics.' *Quarterly Reviews of Biophysics*, **17**, 125.

Edmonds, D. T., Rogers, N. K. & Sternberg, M. J. E. (1984). 'Regular representation of irregular charge distributions. Application to the electrostatic potentials of globular proteins.' *Molecular Physics*, **52**, 1487.

Edsall, J. T. (1968). In *Structural Chemistry and Molecular Biology*, eds A. Rich & N. Davidson, pp. 88–97. San Francisco: Freeman.

Eisenberg, D. & Kauzmann, W. (1969). *The Structure and Properties of Water*. Oxford: Oxford University Press.

Eisenberg, D. & McLachlan, A. D. (1986). 'Solvation energy in protein folding and binding.' *Nature*, **319**, 199.

Eisenberg, E. & Hill, T. L. (1985). 'Muscle contraction and free energy transduction in biological systems.' *Science*, **227**, 999.

Eisenberg, H. (1976). *Biological Macromolecules and Polyelectrolytes in Solution*. Oxford: Clarendon.

Englander, S. W., Downer, N. W. & Teitelbaum, H. (1972). 'Hydrogen exchange.' *Annual Review of Biochemistry*, **41**, 903.

Englander, S. W. & Kallenbach, N. R. (1984). 'Hydrogen exchange and structural dynamics of proteins and nucleic acids.' *Quarterly Reviews of Biophysics*, **16**, 521.

Ermak, D. L. & McCammon, J. A. (1978). 'Brownian dynamics with hydrodynamic interactions.' *Journal of Chemical Physics*, **69**, 1352.

Evans, D. J. & Holian, B. L. (1985). 'The Nose–Hoover thermostat.' *Journal of Chemical Physics*, **83**, 4069.

Ewen, B. (1984). 'Segmental dynamics of chain molecules in polymer solutions and melts as studied by quasielastic neutron scattering.' *Pure & Applied Chemistry*, **56**, 1407.

Feigon, J., Wang, A. H. J., van der Marel, G. A., van Boom, J. H. & Rich, A. (1984). 'A one- and two-dimensional NMR study of the B to Z transition of $(m^5dC\text{-}dG)_3$ in methanolic solution.' *Nucleic Acids Research*, **12**, 1243.

Fersht, A. (1985). *Enzyme Structure and Mechanism*, 2nd edn, San Francisco: Freeman.

Fersht, A. R., Shi, J. P., Wilkinson, A. J., Blow, D. M., Carter, P., Waye, M. Y. & Winter, G. P. (1984). 'Analysis of enzyme structure and activity by protein engineering.' *Angewandte Chemie, International Edition in English*, **23**, 467.

Fersht, A. R., Shi, J. P., Knill-Jones, J., Lowe, D. M., Wilkinson, A. J., Blow, D. M., Brick, P., Carter, P., Waye, M. M. Y. & Winter, G. (1985). 'Hydrogen bonding and biological specificity analysed by protein engineering.' *Nature*, **314**, 235.

Fiamingo, F. G., Thorkildsen, R. & Brill, A. S. (1980). 'Structural distribution and rotational disorder in myoglobin crystals.' *Biophysical Journal*, **32**, 634.

Finzel, B. C. & Salemme, F. R. (1985). 'Lattice mobility and anomalous temperature factor behaviour in cytochrome c′.' *Nature*, **315**, 686.

Fixman, M. (1982). 'The flexibility of polyelectrolyte molecules.' *Journal of Chemical Physics*, **76**, 6346.

Fixman, M. (1986). 'Implicit algorithm for Brownian dynamics of polymers.' *Journal of Chemical Physics*, in press.

Fletcher, R. (1972). 'Conjugate direction methods.' In *Numerical Methods for Unconstrained Optimization*, pp. 73–87. London: Academic.

Fletcher, R. (1980). *Practical Methods of Optimization. Volume 1. Unconstrained Optimization*. New York: Wiley.

Fletcher, R. & Reeves, C. M. (1964). 'Function minimization by conjugate gradients.' *Computer Journal*, **7**, 149.

Flory, P. J. (1969). *Statistical Mechanics of Chain Molecules*. New York: Wiley.

Frank-Kamenetskii, M. D., Lukashin, A. V. & Anshelevich, V. V. (1985). 'Application of polyelectrolyte theory to the study of the B–Z transition in DNA.' *Journal of Biomolecular Structure and Dynamics*, **3**, 35.

Franks, F. (1983). *Water*. London: The Royal Society of Chemistry.

Fratini, A. V., Kopka, M. L., Drew, H. R. & Dickerson, R. E. (1982). 'Reversible bending and helix geometry in a B-DNA dodecamer: CGCGAATT^{Br}CGCG.' *Journal of Biological Chemistry*, **257**, 14686.

Frauenfelder, H., Petsko, G. A. & Tsernoglou, D. (1979). 'Temperature-dependent X-ray diffraction as a probe of protein structural dynamics.' *Nature*, **280**, 558.

Frauenfelder, H. & Wolynes, P. G. (1985). 'Rate theories and puzzles of hemeprotein kinetics.' *Science*, **229**, 337.

Frederick, C. A., Grable, J., Melia, M., Samudzi, C., Jen-Jacobson, L., Wang, B. C., Greene, P., Boyer, H. W. & Rosenberg, J. M. (1984). 'Kinked DNA in crystalline complex with EcoRI endonuclease.' *Nature*, **309**, 327.

Friedman, H. L. (1981). 'Electrolyte solutions at equilibrium.' *Annual Review of Physical Chemistry*, **32**, 179.

Friedman, H. L. (1985*a*). *A Course in Statistical Mechanics*. Englewood Cliffs, New Jersey: Prentice-Hall.

Friedman, H. L. (1986). 'Methods to determine structure in water and aqueous solutions.' *Methods in Enzymology*: *Biomembranes*, **127**, 3.

Friedman, J. M. (1985*b*). 'Structure, dynamics, and reactivity in hemoglobin. *Science*, **228**, 1273.

Friesner, R. A. & Levy, R. M. (1986). 'Theoretical methods for studying collective modes in proteins.' In *Biological Applications of Resonance Raman Spectroscopy*, vol. I, ed. T. Spiro, in press.

Gallion, S. L., Levy, R. M., Weiner, P. K. & Hirata, F. (1986). 'Implementation of a macromolecular mechanics program on a CYBER 205 supercomputer.' *Computers & Chemistry*, **10**, 164.

Ganti, G., Allison, S. A. & McCammon, J. A. (1985). 'Brownian dynamics of diffusion-controlled reactions: The lattice method.' *Journal of Physical Chemistry*, **89**, 3899.

Ganti, G. & McCammon, J. A. (1986). 'Transport properties of macromolecules by Brownian dynamics simulation: Vectorization of Brownian dynamics on the CYBER 205.' *Journal of Computational Chemistry*, **7**, 457.

García de la Torre, J. & Bloomfield, V. A. (1981). 'Hydrodynamic properties of complex, rigid, biological macromolecules: Theory and applications.' *Quarterly Reviews of Biophysics*, **14**, 81.

Garrity, D. K. & Skinner, J. L. (1983). 'Effect of potential shape on isomerization rate constants for the BGK model.' *Chemical Physics Letters*, **95**, 46.

Gear, C. W. (1971). *Numerical Initial Value Problems in Ordinary Differential Equations*. New York: Prentice-Hall.

Geiger, A. (1981). 'Molecular dynamics simulation study of the negative hydration effect in aqueous electrolyte solutions. *Berichte der Bunsen Gesellschaft für Physikalische Chemie*, **85**, 52.

Gelin, B. (1976). 'Application of Empirical Energy Functions to Conformational Problems in Biochemical Systems.' Thesis, Harvard University.

Gelin, B. R. & Karplus, M. (1975). 'Sidechain torsional potentials and motion of amino acids in proteins: Bovine pancreatic trypsin inhibitor.' *Proceedings of the National Academy of Sciences of the USA*, **72**, 2002.

Genzel, L. F., Keilmann, F., Martin, T. P., Winterling, G., Yacoby, Y., Fröhlich, H. & Makinen, M. W. (1976). 'Low-frequency Raman spectra of lysozyme.' *Biopolymers*, **15**, 219.

Germond, J. E., Hirt, B., Oudet, P., Gross-Bellard, M. & Chambon, P. (1975). 'Folding of the DNA double helix in chromatin-like structures from simian virus 40.' *Proceedings of the National Academy of Sciences of the USA*, **72**, 1843.

Getzoff, E. D. (1985). 'Electrostatic recognition between antigen and antibody.' In *Molecular Dynamics and Protein Structure*, ed. J. Hermans, pp. 115–18. Chapel Hill: University of North Carolina.

Getzoff, E. D., Tainer, J. A., Weiner, P. K., Kollman, P. A., Richardson, J. S. & Richardson, D. C. (1983). 'Electrostatic recognition between superoxide and copper, zinc superoxide dismutase.' *Nature*, **306**, 287.

Gibson, Q. H., Olson, J. S., McKinnie, R. E. & Rohlfs, R. J. (1986). 'A kinetic description of ligand binding to sperm whale myoglobin.' *Journal of Biological Chemistry*, **261**, 10228.

Gilson, M. K., Rashin, A., Fine, R. & Honig, B. (1985). 'On the calculation of electrostatic interactions in proteins.' *Journal of Molecular Biology*, **183**, 503.

Glasstone, S., Laidler, K. J. & Eyring, H. (1941). *The Theory of Rate Processes*. New York: McGraw-Hill.

Glover, I., Haneef, I., Pitts, J., Wood, S., Moss, D., Tickle, I. & Blundell, T. (1983). 'Conformational flexibility in a small globular hormone: X-ray analysis of avian pancreatic polypeptide at 0.98-Å resolution.' *Biopolymers*, **22**, 293.

Gō, N., Noguti, T. & Nishikawa, T. (1983). 'Dynamics of a small globular protein, in terms of low-frequency vibrational modes.' *Proceedings of the National Academy of Sciences of the USA*, **80**, 3696.

Goldanskii, V. I., Krupyanskii, Y. F. & Flerov, V. N. (1984). 'Tunneling between quasi-degenerate conformational states and low-temperature heat capacity of biopolymers.' *Doklady Biophysics*, **272**, 209.

Goldstein, H. (1980). *Classical Mechanics*, 2nd edn. Reading, Massachusetts: Addison-Wesley.

Griffey, R. H., Redfield, A. G., Loomis, R. E. & Dahlquist, F. W. (1985). 'Nuclear magnetic resonance observation and dynamics of specific amide protons in T4 lysozyme.' *Biochemistry*, **24**, 817.

Grote, R. F. & Hynes, J. T. (1980). 'The stable states picture of chemical reactions. II. Rate constants for condensed and gas phase reaction models.' *Journal of Chemical Physics*, **73**, 2715.

Guillot, B. & Guissani, Y. (1985). 'Investigation of the chemical potential by molecular dynamics simulation.' *Molecular Physics*, **54**, 455.

Gurd, F. R. N. & Rothgeb, T. M. (1979). 'Motions in proteins.' *Advances in Protein Chemistry*, **33**, 73.

Hagler, A. T. & Moult, J. (1978). 'Computer simulation of the solvent structure around biological macromolecules.' *Nature*, **272**, 222.
Hagler, A. T., Osguthorpe, D. J., Dauber-Osguthorpe P. & Hempel, J. C. (1985). 'Dynamics and conformational energetics of a peptide hormone: Vasopressin.' *Science*, **227**, 1309.
Hagler, A. T., Osguthorpe, D. J. & Robson, B. (1980). 'Monte Carlo simulation of water behavior around the dipeptide N-acetylalanyl-N-methylamide.' *Science*, **208**, 599.
Haile, J. M. & Graben, H. W. (1980). 'Molecular dynamics simulations extended to various ensembles. 1. Equilibrium properties in the isoenthalpic–isobaric ensemble.' *Journal of Chemical Physics*, **73**, 2412.
Hanggi, P. (1983). 'Physics of ligand migration in biomolecules'. *Journal of Statistical Physics*, **30**, 401.
Hanson, D. C., Yguerabide, J. & Schumaker, V. N. (1985). 'Rotational dynamics of immunoglobulin G antibodies anchored in protein A soluble complexes.' *Molecular Immunology*, **22**, 237.
Haran, T. E., Berkovich-Yellin, Z. & Shakked, Z. (1984). 'Base-stacking interactions in double-helical DNA structures: Experiment versus theory.' *Journal of Biomolecular Structure and Dynamics*, **2**, 397.
Harrington, W. F. (1971). 'A mechanochemical mechanism for muscle contraction.' *Proceedings of the National Academy of Sciences of the USA*, **68**, 685.
Harrison, R. W. (1984). 'Variational calculation of the normal modes of a large macromolecule: Methods and some initial results.' *Biopolymers*, **23**, 2943.
Harrison, S. C. (1978). 'Structure of simple viruses: Specificity and flexibility in protein assemblies.' *Trends in Biochemical Sciences*, **3**, 3.
Hartmann, H., Parak, F., Steigemann, W., Petsko, G. A., Ringe Ponzi, D. & Frauenfelder, H. (1982). 'Conformational substates in a protein: Structure and dynamics of metmyoglobin at 80 K.' *Proceedings of the National Academy of Sciences of the USA*, **79**, 4967.
Harvey, S. C. (1979). 'Transport properties of particles with segmental flexibility. I. Hydrodynamic resistance and diffusion coefficients of a freely hinged particle.' *Biopolymers*, **18**, 1081.
Harvey, S. C. (1983). 'DNA structural dynamics: Longitudinal breathing as a possible mechanism for the B⇌Z transition.' *Nucleic Acids Research*, **11**, 4867.
Harvey, S. C. (1986). 'Conformational dynamics of transfer RNA.' *Comments on Molecular & Cellular Biophysics*, **3**, 219.
Harvey, S. C. & Cheung, H. C. (1972). 'Computer simulation of fluorescence depolarization due to Brownian motion.' *Proceedings of the National Academy of Sciences of the USA*, **69**, 3670.
Harvey, S. C. & Cheung, H. C. (1977). 'Fluorescence depolarization studies on the flexibility of myosin rod.' *Biochemistry*, **16**, 5181.
Harvey, S. C. & Cheung, H. C. (1980). 'Transport properties of particles with segmental flexibility. II. Decay of fluorescence polarization anisotropy from hinged macromolecules.' *Biopolymers*, **19**, 913.
Harvey, S. C. & Cheung, H. C. (1982). 'Myosin flexibility'. In *Cell and Muscle*

Motility, vol. 2, eds R. M. Dowben & J. W. Shay, pp. 279–302. New York: Plenum.

Harvey, S. C. & McCammon, J. A. (1981). 'Intramolecular flexibility in phenylalanine transfer RNA.' *Nature*, **294**, 286.

Harvey, S. C. & McCammon, J. A. (1982). 'Macromolecular conformational energy minimization: An algorithm varying pseudodihedral angles.' *Computers & Chemistry*, **6**, 173.

Harvey, S. C., Mellado, P. & García de la Torre, J. (1983). 'Hydrodynamic resistance and diffusion coefficients of segmentally flexible macromolecules with two subunits.' *Journal of Chemical Physics*, **78**, 2081.

Harvey, S. C. & Prabhakaran, M. (1986). 'Ribose puckering: Structure, dynamics, energetics and the pseudorotation cycle.' *Journal of the American Chemical Society*, **108**, 6128.

Harvey, S. C., Prabhakaran, M., Mao, B. & McCammon, J. A. (1984). 'Phenylalanine transfer RNA: Molecular dynamics simulation.' *Science*, **223**, 1189.

Harvey, S. C., Prabhakaran, M. & McCammon, J. A. (1985*a*). 'Molecular-dynamics simulation of phenylalanine transfer RNA. I. Methods and general results.' *Biopolymers*, **24**, 1169.

Harvey, S. C., Prabhakaran, M., Suddath, F. L. & McCammon, J. A. (1985*b*). 'Computer graphics and moving pictures in the analysis of intramolecular motions in phenylalanine transfer RNA': In *Molecular Dynamics and Protein Structure*, ed. J. Hermans, p. 161. Chapel Hill: University of North Carolina.

Heinzinger, K. (1985). 'The structure of aqueous electrolyte solutions as derived from MD (molecular dynamics) simulations.' *Pure & Applied Chemistry*, **57**, 1031.

Helfand, E. (1984). 'Dynamics of conformational transitions in polymers.' *Science*, **226**, 647.

Henry, E. R., Levitt, M. & Eaton, W. A. (1985). 'Molecular dynamics simulation of photodissociation of carbon monoxide from hemoglobin.' *Proceedings of the National Academy of Sciences of the USA*, **82**, 2034.

Henry, E. R., Sommer, J. H., Hofrichter, J. & Eaton, W. A. (1983). 'Geminate recombination of carbon monoxide to myoglobin.' *Journal of Molecular Biology*, **166**, 443.

Hermans, J., Berendsen, H. J. C., van Gunsteren, W. F. & Postma, J. P. M. (1984). 'A consistent empirical potential for water-protein interactions.' *Biopolymers*, **23**, 1513.

Hestenes, M. & Stieffel, E. (1952). *Method of Conjugate Gradients for Solving Linear Systems*. Report 1659. Washington D.C.: National Bureau of Standards.

Hilhorst, H. J., Bakker, A. F., Bruin, C., Compagner, A. & Hoogland, A. (1984). 'Special purpose computers in physics.' *Journal of Statistical Physics*, **34**, 987.

Hilinski, E. F. & Rentzepis, P. M. (1983). 'Biological applications of picosecond spectroscopy.' *Nature*, **302**, 481.

Hilton, B. D. & Woodward, C. K. (1978). 'Nuclear magnetic resonance measurement of hydrogen exchange kinetics of single protons in basic pancreatic trypsin inhibitor.' *Biochemistry*, **17**, 3325.

Hingerty, B., Brown, R. S. & Jack, A. (1978). 'Further refinement of the structure of yeast $tRNA^{Phe}$.' *Journal of Molecular Biology*, **124**, 523.

Hoch, J. C., Dobson, C. M. & Karplus, M. (1985). 'Vicinal coupling constants and protein dynamics.' *Biochemistry*, **24**, 3831.

Hockney, R. W. & Jesshope, C. R. (1983). *Parallel Computers*. London: Adam Hilger.

Hogan, M., Le Grange, J. & Austin, B. (1983). 'Dependence of DNA helix flexibility on base composition.' *Nature*, **304**, 752.

Hol, W. G. J. (1985). 'The role of the α-helix dipole in protein function and structure.' *Progress in Biophysics and Molecular Biology*, **45**, 149.

Holbrook, S. R. & Kim, S. H. (1984). Local mobility of nucleic acids as determined from crystallographic data. I. RNA and B form DNA.' *Journal of Molecular Biology*, **173**, 361.

Holley, R. W., Apgar, J., Everett, G. A., Madison, J. T., Marquisee, M., Merrill, S. H., Penswick, J. R. & Zamir, A. (1965). 'Structure of a ribonucleic acid.' *Science*, **147**, 1462.

Holowka, D. A. & Cathou, R. E. (1976). 'Conformation of immunoglobulin M. 2. Nanosecond fluorescence depolarization analysis of segmental flexibility in anti-ε-1-dimethylamino-5-naphthalenesulfonyl-L-lysine anti-immunoglobulin from horse, pig, and shark.' *Biochemistry*, **15**, 3379.

Hoogstein, K. (1963). 'The crystal and molecular structure of a hydrogen-bonded complex between 1-methylthymine and 9-methyladenine.' *Acta Crystallographica*, **16**, 907.

Horn, A. S. & De Ranter, C. J., eds. (1984). *X-ray Crystallography and Drug Action*. Oxford: Oxford University Press.

Huber, R. & Bennett, W. S. (1983). 'Functional significance of flexibility in proteins.' *Biopolymers*, **22**, 261.

Huxley, H. E. (1969). 'The mechanism of muscular contraction.' *Science*, **164**, 1356.

Hynes, J. T. (1977). 'Statistical mechanics of molecular motion in dense fluids.' *Annual Review of Physical Chemistry*, **28**, 301.

Hynes, J. T. (1985). 'Chemical reaction dynamics in solution'. *Annual Review of Physical Chemistry*, **36**, 573.

Ichiye, T. & Karplus, M. (1983). 'Fluorescence depolarization of tryptophan residues in proteins: A molecular dynamics study.' *Biochemistry*, **22**, 2884.

Imoto, T., Johnson, L. N., North, A. C. T., Phillips, D. C. & Rupley, J. A. (1972). 'Vertebrate lysozymes.' In *The Enzymes*, vol. 7, ed. P. D. Boyer. New York: Academic.

Impey, R. W., Madden, P. A. & McDonald, I. R. (1983). 'Hydration and mobility of ions in solution.' *Journal of Physical Chemistry*, **87**, 5071.

Irikura, K. K., Tidor, B., Brooks, B. R. & Karplus, M. (1985). 'Transition from B to Z DNA: Contribution of internal fluctuations to the configurational entropy difference.' *Science*, **229**, 571.

Jack, A., Ladner, J. E. & Klug, A. (1976). 'Crystallographic refinement of yeast

phenylalanine transfer RNA at 2.5 Å resolution.' *Journal of Molecular Biology*, **108**, 619.
Jack, A. & Levitt, M. (1978). 'Refinement of large structures by simultaneous minimization of energy and R factor.' *Acta Crystallographica*, **A34**, 931.
Jacoby, S. L. S., Kowalik, J. S. & Pizzo, J. T. (1972). *Iterative Methods for Nonlinear Optimization Problems*. Englewood Cliffs, New Jersey: Prentice Hall.
Jacrot, B., Cusack, S., Dianoux, A. J. & Engelman, D. M. (1982). 'Inelastic neutron scattering analysis of hexokinase dynamics and its modification on binding of glucose.' *Nature*, **300**, 84.
James, M. N. G. & Sielecki, A. R. (1983). 'Structure and refinement of penicillopepsin at 1.8 Å resolution.' *Journal of Molecular Biology*, **163**, 299.
Janin, J. & Wodak, S. J. (1983). 'Structural domains in proteins and their role in the dynamics of protein function.' *Progress in Biophysics and Molecular Biology*, **42**, 21.
Jardetzky, O. (1981). 'NMR studies of macromolecular dynamics'. *Accounts of Chemical Research*, **14**, 291.
Jeffrey, G. A. (1984). 'Hydrate inclusion compounds.' In *Inclusion Compounds*, vol. 1, eds. J. L. Atwood, J. E. D. Davies & D. D. MacNicol, p. 135. London: Academic Press.
Johnson, L. N. (1984). 'Enzyme-substrate interactions.' In *Inclusion Compounds*, vol. 3, eds. J. L. Atwood, J. E. D. Davies & D. D. MacNicol, pp. 509–69. London: Academic Press.
Johnston, P. D. & Redfield, A. G. (1979). Proton FT NMR studies of tRNA structure and dynamics.' In *Transfer RNA: Structure, Properties, and Recognition*, eds P. R. Schimmel, D. Söll & J. N. Abelson, pp. 191–206. New York: Cold Spring Harbor.
Jorgensen, W. L. (1983). 'Theoretical studies of medium effects on conformational equilibria.' *Journal of Physical Chemistry*, **87**, 5304.
Jorgensen, W. L., Chandrasekhar, J., Buckner, J. K. & Madura, J. D. (1986). 'Computer simulations of organic reactions in solution.' *Annals of the New York Academy of Sciences*, in press.
Jorgensen, W. L., Gao, J. & Ravimohan, C. (1985). 'Monte Carlo simulations of alkanes in water: Hydration numbers and the hydrophobic effect.' *Journal of Physical Chemistry*, **89**, 3470.
Jorgensen, W. L. & Ravimohan, C. (1985). 'Monte Carlo simulation of differences in free energies of hydration.' *Journal of Chemical Physics*, **83**, 3050.
Jorgensen, W. L. & Swensen, C. J. (1985). 'Optimized intermolecular potential functions for amides and peptides. Hydration of amides.' *Journal of the American Chemical Society*, **107**, 1489.
Kabsch, W., Sander, C. & Trifonov, E. N. (1982). 'The ten helical twist angles of B-DNA.' *Nucleic Acids Research*, **10**, 1097.
Kaiser, E. T. & Lawrence, D. S. (1984). 'Chemical mutation of enzyme active sites.' *Science*, **226**, 505.
Kaptein, R., Zuiderweg, E. R. P., Scheek, R. M., Boelens, R. & van Gunsteren, W. F. (1985). 'A protein structure from nuclear magnetic resonance data: lac repressor headpiece.' *Journal of Molecular Biology*, **182**, 179.

Karim, O. A. & McCammon, J. A. (1986). 'Dynamics of a sodium chloride ion pair in water.' *Journal of the American Chemical Society*, **108**, 1762.
Karplus, M. & Kushick, J. N. (1981). 'Method for estimating the configurational entropy of macromolecules.' *Macromolecules*, **14**, 325.
Karplus, M. & McCammon, J. A. (1981*a*). 'The internal dynamics of globular proteins.' *CRC Critical Reviews of Biochemistry*, **9**, 293.
Karplus, M. & McCammon, J. A. (1981*b*). 'Pressure dependence of aromatic ring rotations in proteins: A collisional interpretation.' *Federation of European Biochemical Societies Letters*, **131**, 34.
Karplus, M. & McCammon, J. A. (1983). 'Dynamics of proteins: Elements and function.' *Annual Review of Biochemistry*, **52**, 263.
Karplus, M. & McCammon, J. A. (1986). 'Dynamics of proteins'. *Scientific American*, April, p. 42.
Kasprzak, A. & Weber, G. (1982). 'Fluorescence depolarization and rotational modes of tyrosine in bovine pancreatic trypsin inhibitor.' *Biochemistry*, **21**, 5924.
Keepers, J., Kollman, P. A. & James, T. L. (1984). 'Molecular mechanical studies of base-pair opening in d(CGCGC):d(GCGCG), $dG_5.dC_5$, d(TATAT):d(ATATA), and $dA_5.dT_5$ in the B and Z forms of DNA.' *Biopolymers*, **23**, 2499.
Keepers, J. W., Kollman, P. A., Weiner, P. K. & James, T. L. (1982). 'Molecular mechanical studies of DNA flexibility: Coupled backbone torsion angles and base-pair openings.' *Proceedings of the National Academy of Sciences of the USA*, **79**, 5537.
Keizer, J. (1985). 'Theory of rapid bimolecular reactions in solution and membranes.' *Accounts of Chemical Research*, **18**, 235.
Keller, W. (1975). 'Determination of the number of superhelical turns in simian virus 40 DNA by gel electrophoresis.' *Proceedings of the National Academy of Sciences of the USA*, **72**, 4876.
Kennard, O. (1985). 'Structural studies of DNA fragments: The G.T wobble base pair in A, B and Z DNA; the G.A base pair in B-DNA.' *Journal of Biomolecular Structure and Dynamics*, **3**, 205.
Kim, K. S. & Clementi, E. (1985). 'Energetics and pattern analysis of crystals of proflavine deoxydinucleoside phosphate complex.' *Journal of the American Chemical Society*, **107**, 227.
Kirkwood, J. G. & Riseman, J. (1948). 'The intrinsic viscosities and diffusion constants of flexible macromolecules in solution.' *Journal of Chemical Physics*, **16**, 565.
Kitamura, K., Mizuno, H., Amisaki, T., Tomita, K. I. & Baba, Y. (1984). 'Locally oscillatory motion of RNA helix derived from linear relationships of backbone torsion angles.' *Biopolymers*, **23**, 1169.
Knapp, E. W., Fischer, S. F. & Parak, F. (1982). 'Protein dynamics from Mössbauer spectra. The temperature dependence.' *Journal of Physical Chemistry*, **86**, 5042.
Knowles, J. R. & Albery, W. J. (1977). 'Perfection in enzyme catalysis: The energetics of triosephosphate isomerase.' *Accounts of Chemical Research*, **10**, 105.
Kollman, P. (1985). 'Theory of complex molecular interactions: Computer

graphics, distance geometry, molecular mechanics, and quantum mechanics'. *Accounts of Chemical Research*, **18**, 105.

Kollman, P., Keepers, J. W. & Weiner, P. (1982). 'Molecular-mechanics studies on d(CGCGAATTCGCG)$_2$ and dA$_{12}$.dT$_{12}$: An illustration of the coupling between sugar repuckering and DNA twisting.' *Biopolymers*, **21**, 2345.

Kollman, P., Singh, U. C., Lybrand, T., Rao, S. & Caldwell, J. (1985). 'A perspective on molecular mechanical and molecular dynamical studies involving DNA.' In *Molecular Basis of Cancer, Part A: Macromolecular Structure, Carcinogens, & Oncogenes*, ed. R. Rein, pp. 173–85. New York: Alan R. Liss, Inc.

Konnert, J. H. & Hendrickson, W. A. (1980). 'A restrained-parameter thermal-factor refinement procedure.' *Acta Crystallographica*, **A36**, 344.

Kopka, M. L., Fratini, A. V., Drew, H. R. & Dickerson, R. E. (1983). 'Ordered water structure around a B-DNA dodecamer. A quantitative study.' *Journal of Molecular Biology*, **163**, 129.

Kornberg, R. D. (1974). 'Chromatin structure: A repeating unit of histones and DNA.' *Science*, **184**, 868.

Koshland, D. E. (1963). 'The role of flexibility in enzyme action.' *Cold Spring Harbor Symposia on Quantitative Biology*, **28**, 473.

Kossiakoff, A. A. (1982). 'Protein dynamics investigated by the neutron diffraction–hydrogen exchange technique.' *Nature*, **296**, 713.

Kossiakoff, A. A. (1985). 'The application of neutron crystallography to the study of dynamic and hydration properties of proteins.' *Annual Review of Biochemistry*, **54**, 1195.

Kramers, H. A. (1940). 'Brownian motion in a field of force and the diffusion model of chemical reactions.' *Physica*, **7**, 284.

Krüger, P., Strassburger, W., Wollmer, A. & van Gunsteren, W. F. (1985). 'A comparison of the structure and dynamics of avian pancreatic polypeptide hormone in solution and in the crystal.' *European Biophysics Journal*, **13**, 77.

Kuriyan, J., Petsko, G. A., Levy, R. M. & Karplus, M. (1986). 'The effects of anisotropy and anharmonicity on protein crystallographic refinement: An evaluation by molecular dynamics.' *Journal of Molecular Biology*, **190**, 227.

Ladner, J. E., Jack, A., Robertus, J. D., Brown, R. S., Rhodes, D., Clark, B. F. C. & Klug, A. (1975). 'Structure of yeast phenylalanine transfer RNA at 2.5 Å resolution.' *Proceedings of the National Academy of Sciences of the USA*, **72**, 4414.

Lakowicz, J. R. & Maliwal, B. P. (1983). Oxygen quenching and fluorescence depolarization of tyrosine residues in proteins.' *Journal of Biological Chemistry*, **258**, 4794.

Lakowicz, J. R. & Weber, G. (1973). Quenching of protein fluorescence by oxygen. Detection of structural fluctuations in proteins on the nanosecond time scale.' *Biochemistry*, **12**, 4171.

Lamm, G. (1984). 'Extended Brownian dynamics. III. Three-dimensional diffusion.' *Journal of Chemical Physics*, **80**, 2845.

Leadbetter, A. J. & Lechner, R. E. (1979). In *The Plastically Crystalline State*, ed. J. N. Sherwood, pp. 285–320. New York: Wiley.

LeBret, M. (1978). 'Relationship between the energy of superhelix formation, the shear modulus, and the torsional Brownian motion of DNA,' *Biopolymers*, **17**, 1939.

LeBret, M. & Zimm, B. H. (1984). 'Distribution of counterions around a cylindrical polyelectrolyte and Manning's condensation theory.' *Biopolymers*, **23**, 287.

Lee, C. Y., McCammon, J. A. & Rossky, P. J. (1984). 'The structure of liquid water at an extended hydrophobic surface.' *Journal of Chemical Physics*, **80**, 4448.

Levinthal, C., Fine, R. & Dimmler, G. (1985). 'FASTRUN–A special-purpose hard-wired computing device for molecular mechanics.' In *Molecular Dynamics and Protein Structure*, ed. J. Hermans, pp. 126–9. Chapel Hill: University of North Carolina.

Levitt, M. (1974). 'Energy refinement of hen egg-white lysozyme'. *Journal of Molecular Biology*, **82**, 393.

Levitt, M. (1976). 'A simplified representation of protein conformations for rapid simulation of protein folding.' *Journal of Molecular Biology*, **104**, 59.

Levitt, M. (1978). 'How many base-pairs per turn does DNA have in solution and in chromatin? Some theoretical calculations.' *Proceedings of the National Academy of Sciences of the USA*, **75**, 640.

Levitt, M. (1981). 'Molecular dynamics of hydrogen bonds in bovine pancreatic trypsin inhibitor protein.' *Nature*, **294**, 379.

Levitt, M. (1982). 'Protein conformation, dynamics, and folding by computer simulation.' *Annual Review of Biophysics and Bioengineering*, **11**, 251.

Levitt, M. (1983*a*). 'Molecular dynamics of native protein. I. Computer simulation of trajectories.' *Journal of Molecular Biology*, **168**, 595.

Levitt, M. (1983*b*). 'Molecular dynamics of native protein. II. Analysis and nature of motion.' *Journal of Molecular Biology*, **168**, 621.

Levitt, M. (1983*c*). 'Protein folding by restrained energy minimization and molecular dynamics.' *Journal of Molecular Biology*, **170**, 723.

Levitt, M. (1983*d*). 'Computer simulation of DNA double-helix dynamics.' *Cold Spring Harbor Symposia on Quantitative Biology*, **47**, 251.

Levitt, M. & Lifson, S. (1969). 'Refinement of protein conformations using a macromolecular energy minimization procedure.' *Journal of Molecular Biology*, **46**, 269.

Levitt, M. & Meirovitch, H. (1983). 'Integrating the equations of motion.' *Journal of Molecular Biology*, **168**, 617.

Levitt, M., Sander, C. & Stern, P. S. (1983). 'The normal modes of a protein: Native bovine pancreatic trypsin inhibitor.' *International Journal of Quantum Chemistry: Quantum Biology Symposium*, **10**, 181.

Levitt, M., Sander, C. & Stern, P. S. (1985). 'Protein normal-mode dynamics: Trypsin inhibitor, crambin, ribonuclease and lysozyme.' *Journal of Molecular Biology*, **181**, 423.

Levitt, M. & Warshel, A. (1978). 'Extreme conformational flexibility of the furanose ring in DNA and RNA.' *Journal of the American Chemical Society*, **100**, 2607.

Levy, R. M. (1986). 'Computer simulations of macromolecular dynamics: Models for vibrational spectroscopy and X-ray refinement.' *Annals of the New York Academy of Sciences*, in press.

Levy, R. M., Karplus, M., Kushick, J. & Perahia, D. (1984*a*). 'Evaluation of the configurational entropy for proteins: Application to molecular dynamics simulations of an α-helix.' *Macromolecules*, **17**, 1370.
Levy, R. M., Karplus, M. & McCammon, J. A. (1979). 'Diffusive Langevin dynamics of model alkanes.' *Chemical Physics Letters*, **65**, 4.
Levy, R. M., Karplus, M., McCammon, J. A. (1981). 'Increase of ^{13}C NMR relaxation times in proteins due to picosecond motional averaging.' *Journal of the American Chemical Society*, **103**, 994.
Levy, R. M. & Keepers, J. W. (1986). 'Computer simulations of protein dynamics: Theory and experiment.' *Comments on Molecular & Cellular Biophysics*, **3**, 273.
Levy, R. M., Perahia, D. & Karplus, M. (1982). 'Molecular dynamics of an α-helical polypeptide: Temperature dependence and deviation from harmonic behavior.' *Proceedings of the National Academy of Sciences of the USA*, **79**, 1346.
Levy, R. M., de la Luz Rojas, O. & Friesner, R. A. (1984*b*). 'Quasi-harmonic method for calculating vibrational spectra from classical simulations on multidimensional anharmonic potential surfaces.' *Journal of Physical Chemistry*, **88**, 4233.
Levy, R. M., Sheridan, R. P., Keepers, J. W., Dubey, G. S., Swaminathan, S. & Karplus, M. (1985). 'Molecular dynamics of myoglobin at 298 K: Results from a 300 picosecond computer simulation.' *Biophysical Journal*, **48**, 509.
Levy, R. M., Srinivasan, A. R., Olson, W. K. & McCammon, J. A. (1984*c*). 'Quasi-harmonic method for studying very low frequency modes in proteins.' *Biopolymers*, **23**, 1099.
Levy, R. M. & Szabo, A. (1982). 'Initial fluorescence depolarization of tyrosines in proteins.' *Journal of the American Chemical Society*, **104**, 2073.
Lifshitz, I. M. (1969). 'Some problems of the statistical theory of biopolymers.' *Soviet Physics JETP*, **28**, 1280.
Lifson, S., Hagler, A. T. & Dauber, P. (1979). 'Consistent force field studies of intermolecular forces in hydrogen-bonded crystals. 1. Carboxylic acids, amides, and the C=O...H–hydrogen bonds.' *Journal of the American Chemical Society*, **101**, 5111.
Lifson, S. & Oppenheim, I. (1960). 'Neighbor interactions and internal rotations in polymer molecules. IV. Solvent effect on internal rotations.' *Journal of Chemical Physics*, **33**, 109.
Linderstrom-Lang, K. U. & Schellman, J. A. (1959). 'Protein structure and enzyme activity.' In *The Enzymes*, vol. 1, eds P. D. Boyer, H. Lardy & K. Myrbäck, p. 443. New York: Academic.
Lipari, G. & Szabo, A. (1982). 'Model-free approach to the interpretation of nuclear magnetic resonance relaxation in macromolecules. 2. Analysis of experimental results.' *Journal of the American Chemical Society*, **104**, 4559.
Lipari, G., Szabo, A. & Levy, R. M. (1982). 'Protein dynamics and NMR relaxation: Comparison of simulations with experiment.' *Nature*, **300**, 197.
Lipscomb, W. N. (1982). 'Acceleration of reactions by enzymes'. *Accounts of Chemical Research*, **15**, 232.
Little, J. W. & Hill, S. A. (1985). 'Deletions within a hinge region of a specific

DNA-binding protein.' *Proceedings of the National Academy of Sciences of the USA*, **82**, 2301.
Lybrand, T. P., Ghosh, I. & McCammon, J. A. (1985). 'Hydration of chloride and bromide anions: Determination of relative free energy by computer simulation.' *Journal of the American Chemical Society*, **107**, 7793.
Lybrand, T. P. & Kollman, P. A. (1985). 'Water–water and water–ion potential functions including terms for many body effects.' *Journal of Chemical Physics*, **83**, 2923.
Lybrand, T. P., McCammon, J. A. & Wipff, G. (1986). 'Theoretical calculation of relative binding affinity in host–guest systems.' *Proceedings of the National Academy of Sciences of the USA*, **83**, 833.
Madura, J. D. & Jorgensen, W. L. (1986). 'Ab initio and Monte Carlo calculations for a nucleophilic addition reaction in the gas phase and in aqueous solution.' *Journal of the American Chemical Society*, **108**, 2517.
Mandal, C., Kallenbach, N. R. & Englander, S. W. (1979). 'Base-pair opening and closing reactions in the double helix. A stopped-flow hydrogen exchange study in poly(rA).poly(rU).' *Journal of Molecular Biology*, **135**, 391.
Manning G. S. (1979). 'Counterion binding in polyelectrolyte theory.' *Accounts of Chemical Research*, **12**, 443.
Mao, B. & McCammon, J. A. (1983). 'Theoretical study of hinge bending in L-arabinose-binding protein.' *Journal of Biological Chemistry*, **258**, 12543.
Mao, B. & McCammon, J. A. (1984). 'Structural study of hinge bending in L-arabinose-binding protein.' *Journal of Biological Chemistry*, **259**, 4964.
Mao, B., Pear, M. R., McCammon, J. A. & Northrup, S. H. (1982*a*). 'Molecular dynamics of ferrocytochrome c: Anharmonicity of atomic displacements.' *Biopolymers*, **21**, 1979.
Mao, B., Pear, M. R., McCammon, J. A. & Quiocho, F. A. (1982*b*). 'Hinge-bending in L-arabinose-binding protein.' *Journal of Biological Chemistry*, **257**, 1131.
Marchese, F. T. & Beveridge, D. L. (1984). 'Pattern recognition approach to the analysis of geometrical features of solvation: Application to the aqueous hydration of Li^+, Na^+, K^+, F^-, and Cl^-.' *Journal of the American Chemical Society*, **106**, 3713.
Margoliash, E. & Bosshard, H. R. (1983). 'Guided by electrostatics, a textbook protein comes of age.' *Trends in Biochemistry*, **8**, 316.
Markley, J. L. & Ulrich, E. L. (1984). 'Detailed analysis of protein structure and function by NMR spectroscopy: Survey of resonance assignments.' *Annual Review of Biophysics and Bioengineering*, **13**, 493.
Matthew, J. B. (1985). 'Electrostatic effects in proteins.' *Annual Review of Biophysics and Biophysical Chemistry*, **14**, 387.
Matthew, J. B., Gurd, F. R. N., Garcia-Moreno, E. B., Flanagan, M. A., March, K. L. & Shire, S. J. (1985). 'pH-dependent processes in proteins.' *CRC Critical Reviews in Biochemistry*, **18**, 91.
Matthew, J. B. & Ohlendorf, D. H. (1985). 'Electrostatic deformation of DNA by a DNA-binding protein.' *Journal of Biological Chemistry*, **260**, 5860.
Matthew, J. B. & Richards, F. M. (1984). 'Differential electrostatic stabilization of A-, B-, and Z-forms of DNA.' *Biopolymers*, **23**, 2743.

Matthew, J. B., Weber, P. C., Salemme, F. R. & Richards, F. M. (1983). 'Electrostatic orientation during electron transfer between flavodoxin and cytochrome c.' *Nature*, **301**, 169.

McCammon, J. A. (1976). 'Molecular dynamics study of the bovine pancreatic trypsin inhibitor.' In *Models for Protein Dynamics*, ed. H. J. C. Berendsen, p. 137. Orsay, France: CECAM, Universite de Paris IX.

McCammon, J. A. (1984). 'Protein dynamics.' *Reports on Progress in Physics*, **47**, 1.

McCammon, J. A. & Deutch, J. M. (1976). 'Frictional properties of nonspherical multisubunit structures. Application to tubules and cylinders.' *Biopolymers*, **15**, 1397.

McCammon, J. A., Deutch, J. M. & Felderhof, B. U. (1975). 'Frictional properties of multisubunit structures.' *Biopolymers*, **14**, 2613.

McCammon, J. A., Gelin, B. R. & Karplus, M. (1977). 'Dynamics of folded proteins.' *Nature*, **267**, 585.

McCammon, J. A., Gelin, B. R., Karplus, M. & Wolynes, P. G. (1976). 'The hinge-bending mode in lysozyme.' *Nature*, **262**, 325.

McCammon, J. A., Karim, O. A., Lybrand, T. P. & Wong, C. F. (1986*a*). 'Ionic association in water: From atoms to enzymes.' *Annals of the New York Academy of Sciences*, in press.

McCammon, J. A. & Karplus, M. (1977). 'Internal motions of antibody molecules.' *Nature*, **268**, 765.

McCammon, J. A. & Karplus, M. (1979). 'Dynamics of activated processes in globular proteins.' *Proceedings of the National Academy of Sciences of the USA*, **76**, 3585.

McCammon, J. A. & Karplus, M. (1980*a*). 'Simulation of protein dynamics.' *Annual Review of Physical Chemistry*, **31**, 29.

McCammon, J. A. & Karplus, M. (1980*b*). 'Dynamics of tyrosine ring rotations in a globular protein.' *Biopolymers*, **19**, 1375.

McCammon, J. A. & Karplus, M. (1983). 'The dynamic picture of protein structure.' *Accounts of Chemical Research*, **16**, 187.

McCammon, J. A., Lee, C. Y. & Northrup, S. H. (1983). 'Side-chain rotational isomerization in proteins: A mechanism involving gating and transient packing defects.' *Journal of the American Chemical Society*, **105**, 2232.

McCammon, J. A., Lybrand, T. P., Allison, S. A. & Northrup, S. H. (1986*c*). 'Ligand binding: New theoretical approaches to molecular recognition.' In *Biomolecular Stereodynamics* III, ed. R. H. Sarma, pp. 227. New York: Adenine.

McCammon, J. A. & Northrup, S. H. (1981). 'Gated binding of ligands to proteins.' *Nature*, **293**, 316.

McCammon, J. A., Northrup, S. H. & Allison, S. A. (1986*b*). 'Diffusional dynamics of ligand-receptor association.' *Journal of Physical Chemistry*, **90**, 3901.

McCammon, J. A., Northrup, S. H., Karplus, M. & Levy, R. M. (1980). 'Helix-coil transitions in a simple polypeptide model.' *Biopolymers*, **19**, 2033.

McCammon, J. A. & Wolynes, P. G. (1977). Nonsteady hydrodynamics of biopolymer motions.' *Journal of Chemical Physics*, **66**, 1452.

McCammon, J. A., Wolynes, P. G. & Karplus, M. (1979). 'Picosecond dynamics of tyrosine side chains in proteins.' *Biochemistry*, **18**, 927.

McQuarrie, D. A. (1976). *Statistical Mechanics*. New York: Harper and Row.

Mendelson, R. A., Morales, M. F. & Botts, J. (1973). 'Segmental flexibility of the S-1 moiety of myosin.' *Biochemistry*, **12**, 2250.

Mezei, M., Mehrotra, P. K. & Beveridge, D. L. (1985). 'Monte Carlo determination of the free energy and internal energy of hydration for the Ala dipeptide at 25 °C.' *Journal of the American Chemical Society*, **107**, 2239.

Middendorf, H. D. (1984). 'Biophysical applications of quasi-elastic and inelastic neutron scattering.' *Annual Review of Biophysics and Bioengineering*, **13**, 425.

Millar, D. P., Robbins, R. J. & Zewail, A. H. (1981). 'Time-resolved spectroscopy of macromolecules: effect of helical structure on the torsional dynamics of DNA and RNA.' *Journal of Chemical Physics*, **74**, 4200.

Miller, K. J. (1979). 'Interactions of molecules with nucleic acids. I. An algorithm to generate nucleic acid structures with an application to the B-DNA structure and a counterclockwise helix.' *Biopolymers*, **18**, 959.

Mills, P., Anderson, C. F. & Record, M. T. (1985). 'Monte Carlo studies of counterion–DNA interactions. Comparison of the radial distribution of counterions with predictions of other polyelectrolyte theories.' *Journal of Physical Chemistry*, **89**, 3984.

Mirau, P. A., Behling, R. W. & Kearns, D. R. (1985). 'Internal motions in B- and Z-form poly(dG-dC)·poly (dG-dC): ^{1}H NMR relaxation studies.' *Biochemistry*, **24**, 6200.

Mirau, P. A. & Kearns, D. R. (1983). '^{1}H NMR relaxation studies of the dynamics of the base paired imino protons in cytosine-containing DNA and RNA duplexes.' In *Structure and Dynamics: Nucleic Acids and Proteins*, eds E. Clementi & R. H. Sarma, p. 227. New York: Adenine Press.

Montgomery, J. A., Chandler, D. & Berne, B. J. (1979). 'Trajectory analysis of a kinetic theory for isomerization dynamics in condensed phases.' *Journal of Chemical Physics*, **70**, 4056.

Moog, R. S., Ediger, M. D., Boxer, S. G. & Fayer, M. D. (1982). 'Viscosity dependence of the rotational reorientation of rhodamine B in mono- and polyalcohols. Picosecond transient grating experiments.' *Journal of Physical Chemistry*, **86**, 4694.

Moras, D., Comarmond, M. B., Fischer, J., Weiss, R., Thierry, J. C., Ebel, J. P. & Giegé, R. (1980). 'Crystal structure of yeast $tRNA^{Asp}$.' *Nature*, **288**, 669.

Morgan, J. D., McCammon, J. A. & Northrup, S. H. (1983). 'Molecular dynamics of ferrocytochrome c: Time dependence of the atomic displacements.' *Biopolymers*, **22**, 1579.

Morozova, T. Y. & Morozov, V. N. (1982). 'Viscoelasticity of protein crystal as a probe of the mechanical properties of a protein molecule. Hen egg-white lysozyme.' *Journal of Molecular Biology*, **157**, 173.

Mossing, M. C. & Record, M. T. (1985). 'Thermodynamic origins of specificity in the lac repressor–operator interaction. Adaptability in the recognition of mutant operator sites.' *Journal of Molecular Biology*, **186**, 295.

Müller-Hill, B. (1983). 'Sequence homology between Lac and Gal repressors and three sugar-binding periplasmic proteins.' *Nature*, **302**, 163.

Murthy, C. S., Bacquet, R. J. & Rossky, P. J. (1985). 'Ionic distributions near polyelectrolytes. A comparison of theoretical approaches.' *Journal of Physical Chemistry*, **89**, 701.

Nadler, W. & Schulten, K. (1984). 'Theory of Mössbauer spectra of proteins fluctuating between conformational substates.' *Proceedings of the National Academy of Sciences of the USA*, **81**, 5719.

Neumann, E. (1981). 'Dynamics of molecular recognition in enzyme-catalyzed reactions.' In *Structural and Functional Aspects of Enzyme Catalysis*, eds H. Eggerer & R. Huber, p. 47. Berlin: Springer Verlag.

Newcomer, M. E., Lewis, B. A. & Quiocho, F. A. (1981). 'The radius of gyration of L-arabinose-binding protein decreases upon binding of ligand.' *Journal of Biological Chemistry*, **256**, 13218.

Nguyen, D. T. & Case, D. A. (1985). 'On finding stationary states on large-molecule potential energy surfaces.' *Journal of Physical Chemistry*, **89**, 4020.

Nilsson, L., Rigler, R. & Laggner, P. (1982). 'Structural variability of tRNA: Small-angle X-ray scattering of the yeast $tRNA^{Phe}$-Escherichia coli $tRNA_2^{Glu}$ complex.' *Proceedings of the National Academy of Sciences of the USA*, **79**, 5891.

Noguti, T. & Gō, N. (1982). 'Collective variable description of small-amplitude conformational fluctuations in a globular protein.' *Nature*, **296**, 776.

Noguti, T. & Gō, N. (1985). 'Efficient Monte Carlo method for simulation of fluctuating conformations of native proteins.' *Biopolymers*, **24**, 527.

Noll, M. (1974). 'Subunit structure of chromatin.' *Nature*, **251**, 249.

Northrup, S. H., Allison, S. A. & McCammon, J. A. (1984). 'Brownian dynamics simulation of diffusion-influenced bimolecular reactions.' *Journal of Chemical Physics*, **80**, 1517.

Northrup, S. H., Curvin, M. S., Allison, S. A. & McCammon, J. A. (1986). 'Optimization of Brownian dynamics methods for diffusion-influenced rate constant calculations.' *Journal of Chemical Physics*, **84**, 2196.

Northrup, S. H. & Hynes, J. T. (1980). 'The stable states picture of chemical reactions. I. Formulation for rate constants and initial condition effects.' *Journal of Chemical Physics*, **73**, 2700.

Northrup, S. H. & McCammon, J. A. (1980*a*). 'Simulation methods for protein structure fluctuations.' *Biopolymers*, **19**, 1001.

Northrup, S. H. & McCammon, J. A. (1980*b*). 'Efficient trajectory simulation methods for diffusional barrier crossing processes.' *Journal of Chemical Physics*, **72**, 4569.

Northrup, S. H. & McCammon, J. A. (1984). 'Gated reactions.' *Journal of the American Chemical Society*, **106**, 930.

Northrup, S. H., Pear, M. R., Lee, C. Y., McCammon, J. A. & Karplus, M. (1982*a*). 'Dynamical theory of activated processes in globular proteins.' *Proceedings of the National Academy of Sciences of the USA*, **79**, 4035.

Northrup, S. H., Pear, M. R., McCammon, J. A., Karplus, M. & Takano, T. (1980). 'Internal mobility of ferrocytochrome c.' *Nature*, **287**, 659.

Northrup, S. H., Pear, M. R., Morgan, J. D., McCammon, J. A. & Karplus, M. (1981). 'Molecular dynamics of ferrocytochrome c. Magnitude and anisotropy of atomic displacements.' *Journal of Molecular Biology*, **153**, 1087.

Northrup, S. H., Smith, J. D., Boles, J. O. & Reynolds, J. C. L. (1986). 'The effect of dipole moment on diffusion-controlled bimolecular reaction rates.' *Journal of Chemical Physics*, **84**, 5536.

Northrup, S. H., Zarrin, F. & McCammon, J. A. (1982*b*). 'Rate theory for gated diffusion-influenced ligand binding to proteins.' *Journal of Physical Chemistry*, **86**, 2314.

Ohlendorf, D. H. & Matthews, B. W. (1983). 'Structural studies of protein–nucleic acid interactions.' *Annual Review of Biophysics and Bioengineering*, **12**, 259.

Olins, A. L. & Olins, D. E. (1974). 'Spheroid chromatin units (ν bodies).' *Science*, **183**, 330.

Olson, T., Fournier, M. J., Langley, K. H. & Ford, N. C. (1976). 'Detection of a major conformational change in transfer ribonucleic acid by laser light scattering.' *Journal of Molecular Biology*, **102**, 193.

Olson, W. K. (1980). 'Configurational statistics of polynucleotide chains. An updated virtual bond model to treat effects of base stacking.' *Macromolecules*, **13**, 721.

Olson, W. K. (1981). 'Understanding the motions of DNA.' *Proceedings of the Second SUNYA Conversation in the Discipline Biomolecular Stereodynamics*, ed. R. H. Sarma, pp. 327–43. New York: Adenine.

Olson, W. K. (1982). 'How flexible is the furanose ring? 2. An updated potential energy estimate.' *Journal of the American Chemical Society*, **104**, 278.

Olson, W. K. & Flory, P. J. (1972). Spatial configurations of polynucleotide chains. I. Steric interactions in polyribonucleotides: A virtual bond model.' *Biopolymers*, **11**, 1.

Olson, W. K., Srinivasan, A. R., Marky, N. L. & Balaji, V. N. (1983). 'Theoretical probes of DNA conformation examining the B→Z conformational transition.' *Cold Spring Harbor Symposia on Quantitative Biology*, **47**, 229.

Olson, W. K. & Sussman, J. L. (1982). 'How flexible is the furanose ring? 1. A comparison of experimental and theoretical studies.' *Journal of the American Chemical Society*, **104**, 270.

Ostlund, N. S. & Whiteside, R. A. (1985). 'A machine architecture for molecular dynamics – The systolic loop.' *Annals of the New York Academy of Sciences*, **439**, 195.

Pabo, C. O. & Sauer, R. T. (1984). 'Protein–DNA recognition.' *Annual Review of Biochemistry*, **53**, 293.

Pack, G. R. & Klein, B. J. (1984). 'Generalized Poisson–Boltzmann calculation of the distribution of electrolyte ions around the B- and Z-conformers of DNA.' *Biopolymers*, **23**, 2801.

Painter, P. C. & Mosher, L. E. (1979). 'The low-frequency Raman spectrum of an antibody molecule: Bovine IgG.' *Biopolymers*, **18**, 3121.

Painter, P. C., Mosher, L. & Rhoads, C. (1981). 'Low-frequency modes in the Raman spectrum of DNA.' *Biopolymers*, **20**, 243.

Pangali, C., Rao, M. & Berne, B. J. (1979). 'A Monte Carlo simulation of the hydrophobic interaction.' *Journal of Chemical Physics*, **71**, 2975.

Parak, F. & Knapp, E. W. (1984). 'A consistent picture of protein dynamics.' *Proceedings of the National Academy of Sciences of the USA*, **81**, 7088.

Parry, D. A. D. & Baker, E. N. (1984). 'Biopolymers.' *Reports on Progress in Physics*, **47**, 1133.

Paterson, Y., Némethy, G. & Scheraga, H. A. (1981). 'Hydration of amino acids, peptides, and model compounds.' *Annals of the New York Academy of Sciences*, **367**, 132.

Pauling, L. (1960). *The Nature of the Chemical Bond*. New York: Cornell University Press.

Pear, M. R., Northrup, S. H. & McCammon, J. A. (1980). 'Diffusional correlations in polymer dynamics.' *Journal of Chemical Physics*, **73**, 4703.

Pear, M. R., Northrup, S. H., McCammon, J. A., Karplus, M. & Levy, R. M. (1981). 'Correlated helix–coil transitions in polypeptides.' *Biopolymers*, **20**, 629.

Perrin, F. (1934). 'Mouvement Brownien d'un ellipsoide (I). Dispersion diélectrique pour des molécules ellipsoidales.' *Le Journal de Physique et Le Radium*, **7**, 497.

Perutz, M. F. (1984). 'Species adaptation in a protein molecule.' *Advances in Protein Chemistry*, **36**, 213.

Perutz, M. F. & Mathews, F. S. (1966). 'An X-ray study of azide methaemoglobin.' *Journal of Molecular Biology*, **21**, 199.

Peticolas, W. L. (1979). 'Low frequency vibrations and the dynamics of proteins and polypeptides.' *Methods in Enzymology*, **61**, 425.

Petsko, G. A. & Ringe, D. (1984). 'Fluctuations in protein structure from X-ray diffraction.' *Annual Review of Biophysics and Bioengineering*, **13**, 331.

Pettitt, B. M. (1986). 'Intramolecular dielectric screening.' *Journal of Chemical Physics*, in press.

Pettitt, B. M. & Karplus, M. (1985). 'The potential of mean force surface for the alanine dipeptide in aqueous solution: A theoretical approach.' *Chemical Physics Letters*, **121**, 194.

Pettitt, B. M. & Karplus, M. (1986). 'Interaction energies: Their role in drug design.' In *Drug Design and Molecular Graphics*, London: Medical Research Council.

Phillips, D. C. (1981). 'Crystallographic studies of movement within proteins.' *Biochemical Society Symposia*, **46**, 1.

Pippard, A. B. (1964). *Elements of Classical Thermodynamics for Advanced Students of Physics*. Cambridge University Press.

Postma, J. P. M., Berendsen, H. J. C. & Haak, J. R. (1982). 'Thermodynamics of cavity formation in water.' *Faraday Symposium of the Chemical Society*, **17**, 55.

Prabhakaran, M. & Harvey, S. C. (1985). 'Molecular dynamics annealing of large deformations of macromolecules: Stretching the DNA double helix to form an intercalation site.' *Journal of Physical Chemistry*, **89**, 5767.

Prabhakaran, M., Harvey, S. C., Mao, B. & McCammon, J. A. (1983). 'Molecular dynamics of phenylalanine transfer RNA.' *Journal of Biomolecular Structure and Dynamics*, **1**, 357.

Prabhakaran, M., Harvey, S. C. & McCammon, J. A. (1985*a*). 'Molecular-dynamics simulation of phenylalanine transfer RNA. II. Amplitudes, anisotropies, and anharmonicities of atomic motions.' *Biopolymers*, **24**, 1189.

Prabhakaran, M., McCammon, J. A. & Harvey, S. C. (1985*b*). 'Atomic motions in phenylalanine transfer RNA probed by molecular dynamics simulations.' In *Molecular Basis of Cancer, Part A: Macromolecular Structure, Carcinogens, and Oncogenes*, ed. R. Rein, pp. 123–9. New York: Alan R. Liss.

Ptitsyn, O. B. (1978). 'Inter-domain mobility in proteins and its probable functional role.' *Federation of European Biochemical Societies Letters*, **93**, 1.

Pulleyblank, D. E., Shure, M., Tang, D., Vinograd, J. & Vosberg, H.-P. (1975). 'Action of nicking-closing enzyme on supercoiled and nonsupercoiled closed circular DNA: Formation of a Boltzmann distribution of topological isomers.' *Proceedings of the National Academy of Sciences of the USA*, **72**, 4280.

Quigley, G. J., Seeman, N. C., Wang, A. H. J., Suddath, F. L. & Rich, A. (1975). 'Yeast phenylalanine transfer RNA: Atomic coordinates and torsion angles.' *Nucleic Acids Research*, **2**, 2329.

Rahman, A. (1964). 'Correlations in the motion of atoms in liquid argon.' *Physical Review*, **136A**, 405.

Rahman, A. (1966). 'Liquid structure and self-diffusion.' *Journal of Chemical Physics*, **45**, 2585.

Rahman, A. & Stillinger, F. H. (1971). 'Molecular dynamics study of liquid water.' *Journal of Chemical Physics*, **55**, 3336.

Ramachandran, G. N., Ramakrishnan, C. & Sasisekharan, V. (1963). 'Stereochemistry of polypeptide chain configurations.' *Journal of Molecular Biology*, **7**, 95.

Ramstein, J. & Leng, M. (1980). 'Salt-dependent dynamic structure of poly(dG–dC). poly(dG–dC).' *Nature*, **288**, 413.

Rao, S. N. & Kollman, P. (1985). 'On the role of uniform and mixed sugar puckers in DNA double helical structures.' *Journal of the American Chemical Society*, **107**, 1611.

Rao, S. T., Westhof, E. & Sundaralingam, M. (1981). 'Exact method for the calculation of pseudorotation parameters P, τ_m and their errors. A comparison of the Altona–Sundaralingam and Cremer–Pople treatment of puckering of five-membered rings.' *Acta Crystallographica*, **A37**, 421.

Ravishanker, G., Mezei, M. & Beveridge, D. L. (1982). 'Monte Carlo computer simulation study of the hydrophobic effect. Potential of mean force for $[(CH_4)_2]_{aq}$ at 25 and 50 °C.' *Faraday Symposium of the Chemical Society*, **17**, 79.

Read, R. J., Brayer, G. D., Jurášek, L. & James, M. N. G. (1984). 'Critical evaluation of comparative model building of streptomyces griseus trypsin.' *Biochemistry*, **23**, 6570.

Record, M. T., Anderson, C. F. & Lohman, T. M. (1978). 'Thermodynamic analysis of ion effects on the binding and conformational equilibria of proteins and nucleic acids: The roles of ion association or release, screening, and ion effects on water activity.' *Quarterly Reviews of Biophysics*, **11**, 103.

Record, M. T., Anderson, C. F., Mills, P., Mossing, M. & Roe, J. H. (1986). 'Ions as regulators of protein–nucleic acid interactions *in vitro* and *in vivo*.' *Advances in Biophysics*, in press.

Reid, B. R. (1977). 'Synthetase–tRNA Recognition.' In *Nucleic Acid–Protein Recognition*, ed. H. J. Vogel, pp. 375–90. New York: Academic.

Rein, R., ed. (1985). *Molecular Basis of Cancer. Part A: Macromolecular Structure, Carcinogens, and Oncogenes*. New York: Alan R. Liss.

Rich, A. (1978). 'Transfer RNA: Three-dimensional structure and biological function.' *Trends in Biochemical Sciences*, **3**, 34.

Richards, F. M. (1977). 'Areas, volumes, packing, and protein structure.' *Annual Review of Biophysics and Bioengineering*, **6**, 151.

Richards, W. G. (1984). 'Quantum Pharmacology.' *Endeavour*, **8**, 172.

Richardson, J. S. (1981). 'The anatomy and taxonomy of protein structure.' *Advances in Protein Chemistry*, **34**, 167.

Richarz, R., Nagayama, K. & Wüthrich, K. (1980). 'Carbon-13 nuclear magnetic resonance relaxation studies of internal mobility of the polypeptide chain in basic pancreatic trypsin inhibitor and a selectively reduced analogue.' *Biochemistry*, **19**, 5189.

Richmond, T. J. (1984). 'Solvent accessible surface area and excluded volume in proteins. Analytical equations for overlapping spheres and implications for the hydrophobic effect.' *Journal of Molecular Biology*, **178**, 63.

Richter, P. H. & Eigen, M. (1974). 'Diffusion controlled reaction rates in spheroidal geometry. Application to repressor–operator association and membrane bound enzymes.' *Biophysical Chemistry*, **2**, 255.

Rigler, R. & Wintermeyer, W. (1983). 'Dynamics of tRNA.' *Annual Review of Biophysics and Biophysical Chemistry*, **12**, 475.

Ringe, D., Petsko, G. A., Kerr, D. E. & Ortiz de Montellano, P. R. (1984). 'Reaction of myoglobin with phenylhydrazine: A molecular doorstop'. *Biochemistry*, **23**, 2.

Ringe, D. & Petsko, G. A. (1985). 'Mapping protein dynamics by X-ray diffraction.' *Progress in Biophysics and Molecular Biology*, **47**, 197.

Robertus, J. D., Ladner, J. E., Finch, J. T., Rhodes, D., Brown, R. S., Clark, B. F. C. & Klug, A. (1974). 'Structure of yeast phenylalanine tRNA at 3 Å resolution.' *Nature*, **250**, 546.

Robinson, B. H., Lerman, L. S., Beth, A. H., Frisch, H. L., Dalton, L. R. & Auer, C. (1980). 'Analysis of double-helix motions with spin-labeled probes: Binding geometry and the limit of torsional elasticity.' *Journal of Molecular Biology*, **139**, 19.

Rogers, N. K. & Sternberg, M. J. E. (1984). 'Electrostatic interactions in globular proteins. Different dielectric models applied to the packing of α-helices.' *Journal of Molecular Biology*, **174**, 527.

Rosenberg, R. O., Berne, B. J. & Chandler, D. (1980). 'Isomerization dynamics in liquids by molecular dynamics.' *Chemical Physics Letters*, **75**, 162.

Rossky, P. J. (1985). 'The structure of polar molecular liquids.' *Annual Review of Physical Chemistry*, **36**, 321.

Rossky, P. J. & Karplus, M. (1979). 'Solvation. A molecular dynamics study of a dipeptide in water.' *Journal of American Chemical Society*, **101**, 1913.

Ryckaert, J. P., Ciccotti, G. & Berendsen, H. J. C. (1977). 'Numerical integration of the Cartesian equations of motion of a system with constraints: Molecular dynamics of n-alkanes.' *Journal of Computational Physics*, **23**, 327.

Saenger, W. (1984). *Principles of Nucleic Acid Structure*. New York: Springer-Verlag.

Sams, C. F., Vyas, N. K., Quiocho, F. A. & Matthews, K. S. (1984). 'Predicted structure of the sugar-binding site of the lac repressor.' *Nature*, **310**, 429.

Sarma, R. H., ed. (1981*a*). *Biomolecular Stereodynamics, Volume* I, New York: Adenine Press.

Sarma, R. H., ed. (1981*b*). *Biomolecular Stereodynamics, Volume* II, New York: Adenine Press.

Schaefer, J., Stejskal, E. O., Perchak, D., Skolnick, J. & Yaris, R. (1985). 'Molecular mechanism of the ring-flip process in polycarbonate.' *Macromolecules*, **18**, 368.

Scheraga, H. A. (1968). 'Calculations of conformations of polypeptides.' *Advances in Physical Organic Chemistry*, **6**, 103.

Schevitz, R. W., Podjarny, A. D., Krishnamachari, N., Hughes, J. J., Sigler, P. B. & Sussman, J. L. (1979). 'Crystal structure of a eukaryotic initiator tRNA.' *Nature*, **278**, 188.

Schimmel, P. R., Söll, D. & Abelson, J. N., eds (1979). *Transfer RNA: Structure, Properties, and Recognition*. New York: Cold Spring Harbor.

Schulz, G. E. & Schirmer, R. H. (1979). *Principles of Protein Structure*. New York: Springer-Verlag.

Schwarz, G. & Engel, J. (1972). 'Kinetics of cooperative conformational transitions of linear biopolymers.' *Angewandte Chemie International Edition*, **11**, 568.

Schwarz, U., Möller, A. & Gassen, H. G. (1978). 'Induced structural transitions in tRNA.' *Federation of European Biochemical Societies Letters*, **90**, 189.

Seibel, G. L., Singh, U. C. & Kollman, P. A. (1986). 'A molecular dynamics simulation of double helical B-DNA including counterions and water.' *Proceedings of the National Academy Sciences of the USA*, in press.

Shakked, Z., Rabinovich, D., Kennard, O., Cruse, W. B. T., Salisbury, S. A. & Viswamitra, M. A. (1983). 'Sequence dependent conformation of an A-DNA double helix. The crystal structure of the octamer d(G-G-T-A-T-A-C-C)'. *Journal of Molecular Biology*, **166**, 183.

Shank, C. V. (1983). 'Measurement of ultrafast phenomena in the femtosecond time domain.' *Science*, **219**, 1027.

Shaw, P. B. (1985). 'Theory of the Poisson Green's function for discontinuous dielectric media with an application to protein biophysics.' *Physical Review A*, **32**, 2476.

Sheriff, S., Hendrickson, W. A., Stenkamp, R. E., Sieker, L. C. & Jensen, L. H. (1985). 'Influence of solvent accessibility and intermolecular contacts on atomic mobilities in hemerythrins.' *Proceedings of the National Academy of Sciences of the USA*, **82**, 1104.

Shih, H. H. L., Brady, J. & Karplus, M. (1985). 'Structure of proteins with single-site mutations: A minimum perturbation approach.' *Proceedings of the National Academy of Sciences of the USA*, **82**, 1697.

Shire, S. J., Hanania, G. I. H. & Gurd, F. R. N. (1974). 'Electrostatic effects in myoglobin. Hydrogen ion equilibria in sperm whale ferrimyoglobin.' *Biochemistry*, **13**, 2967.

Shore, D. & Baldwin, R. L. (1983). 'Energetics of DNA Twisting. I. Relation between twist and cyclization probability.' *Journal of Molecular Biology*, **170**, 957.

Shoup, D. & Szabo, A. (1982). 'Role of diffusion in ligand binding to macromolecules and cell-bound receptors.' *Biophysical Journal*, **40**, 33.

Singh, U. C., Weiner, S. J. & Kollman, P. (1985). 'Molecular dynamics simulation of d(C-G-C-G-A)·d(T-C-G-C-G) with and without "hydrated" counterions.' *Proceedings of the National Academy of Sciences of the USA*, **82**, 755.

Skinner, J. L. & Wolynes, P. G. (1978). 'Relaxation processes and chemical kinetics.' *Journal of Chemical Physics*, **69**, 2143.

Smith, J. C., Cusack, S., Pezzeca, U., Brooks, B., Karplus, M. & Finney, J. L. (1985). 'Protein dynamics studied by inelastic neutron scattering'. In *Fourth Conversation in Biomolecular Stereodynamics*, ed. R. H. Sarma, p. 287. New York: Adenine.

Smith, R. N., Hansch, C., Kim, K. H., Omiya, B., Fukumura, G., Selassie, C. D., Jow, P. Y. C., Blaney, J. M. & Langridge, R. (1982). 'The use of crystallography, graphics, and quantitative structure–activity relationships in the analysis of the papain hydrolysis of X-phenyl hippurates.' *Archives of Biochemistry and Biophysics*, **215**, 319.

Söll, D., Abelson, J. N. & Schimmel, P. R., eds (1980). *Transfer RNA: Biological Aspects*. New York: Cold Spring Harbor.

Soumpasis, D. M. (1984). 'Statistical mechanics of the B → Z transition of DNA: Contribution of diffuse ionic interactions.' *Proceedings of the National Academy of Sciences of the USA*, **81**, 5116.

Stein, D. L. (1985). 'A model of protein conformational substates.' *Proceedings of the National Academy of Sciences of the USA*, **82**, 3670.

Stillinger, F. H. (1980). 'Water revisited.' *Science*, **209**, 451.

Stillinger, F. H. & Rahman, A. (1972). 'Molecular dynamics study of temperature effects on water structure and kinetics.' *Journal of Chemical Physics*, **57**, 1281.

Stillinger, F. H. & Rahman, A. (1974). 'Improved simulation of liquid water by molecular dynamics.' *Journal of Chemical Physics*, **60**, 1545.

Stillinger, F. H. & Weber, T. A. (1984). 'Packing structures and transitions in liquids and solids.' *Science*, **225**, 983.

Strogatz, S. (1982). 'Estimating the torsional rigidity of DNA from supercoiling data.' *Journal of Chemical Physics*, **77**, 580.

Suezaki, Y. & Gō, N. (1975). 'Breathing mode of conformational fluctuations in globular proteins.' *International Journal of Peptide and Protein Research*, **7**, 333.

Sundaralingam, M. & Westhof, E. (1981). 'Structural motifs of the nucleotidyl unit and the handedness of polynucleotide helices.' *International Journal of Quantum Chemistry; Quantum Biology Symposium*, **8**, 287.

Sussman, J. L., Holbrook, S. R., Warrant, R. W., Church, G. M. & Kim, S. H.

(1978). 'Crystal structure of yeast phenylalanine transfer RNA. I. Crystallographic refinement.' *Journal of Molecular Biology*, **123**, 607.
Sussman, J. L. & Trifonov, E. N. (1978). 'Possibility of nonkinked packing of DNA in chromatin.' *Proceedings of the National Academy of Sciences of the USA*, **75**, 103.
Swaminathan, S., Ichiye, T., van Gunsteren, W. & Karplus, M. (1982). 'Time dependence of atomic fluctuations in proteins: Analysis of local and collective motions in bovine pancreatic trypsin inhibitor.' *Biochemistry*, **21**, 5230.
Symon, K. R. (1960). *Mechanics*. Reading, Massachusetts: Addison-Wesley.
Szabo, A. (1984). 'Theory of fluorescence depolarization in macromolecules and membranes.' *Journal of Chemical Physics*, **81**, 150.
Szabo, A., Shoup, D., Northrup, S. H. & McCammon, J. A. (1982). 'Stochastically gated diffusion-influenced reactions.' *Journal of Chemical Physics*, **77**, 4484.
Tainer, J. A., Getzoff, E. D., Paterson, Y., Olson, A. J. & Lerner, R. A. (1985). 'The atomic mobility component of protein antigenicity.' *Annual Review of Immunology*, **3**, 501.
Takusagawa, F. & Berman, H. M. (1983). 'Some new aspects of actinomycin D – nucleic acid binding.' *Cold Spring Harbor Symposia on Quantitative Biology*, **47**, 315.
Tanford, C. & Kirkwood, J. G. (1957). 'Theory of protein titration curves. I. General equations for impenetrable spheres.' *Journal of the American Chemical Society*, **79**, 5333.
Taylor, E. R. & Olson, W. K. (1983). 'Theoretical studies of nucleic acid interactions. 1. Estimates of conformational mobility in intercalated chains.' *Biopolymers*, **22**, 2667.
Teeter, M. M. (1984). 'Water structure of a hydrophobic protein at atomic resolution: Pentagon rings of water molecules in crystals of crambin.' *Proceedings of the National Academy of Sciences of the USA*, **81**, 6014.
Tembe, B. L. & McCammon, J. A. (1984). 'Ligand-receptor interactions.' *Computers & Chemistry*, **8**, 281.
Terner, J. & El-Sayed, M. A. (1985). 'Time-resolved resonance Raman spectroscopy of photobiological and photochemical transients.' *Accounts of Chemical Research*, **18**, 331.
Thomas, J. C., Allison, S. A., Appellof, C. J. & Schurr, J. M. (1980). 'Torsion dynamics and depolarization of fluorescence of linear macromolecules. II. Fluorescence polarization anisotropy measurements on a clean viral ϕ29 DNA.' *Biophysical Chemistry*, **12**, 177.
Thomas, P. G., Russell, A. J. & Fersht, A. R. (1985). 'Tailoring the pH dependence of enzyme catalysis using protein engineering.' *Nature*, **318**, 375.
Tidor, B., Irikura, K. K., Brooks, B. R. & Karplus, M. (1983). 'Dynamics of DNA oligomers.' *Journal of Biomolecular Structure and Dynamics*, **1**, 231.
Torchia, D. A. (1984). 'Solid state NMR studies of protein internal dynamics.' *Annual Review of Biophysics and Bioengineering*, **13**, 125.
Tung, C. S. & Harvey, S. C. (1984). 'A molecular mechanical model to predict the helix twist angles of B-DNA.' *Nucleic Acids Research*, **12**, 3343.

Tung, C. S. & Harvey, S. C. (1986*a*). 'Computer graphics program to reveal the dependence of the gross three-dimensional structure of the B-DNA double helix on primary structure.' *Nucleic Acids Research*, **14**, 381.

Tung, C. S. & Harvey, S. C. (1986*b*). 'Base sequence, local helix structure and macroscopic curvature of A-DNA and B-DNA.' *Journal of Biological Chemistry*, **261**, 3700.

Tung, C. S., Harvey, S. C. & McCammon, J. A. (1984). 'Large-amplitude bending motions in phenylalanine transfer RNA.' *Biopolymers*, **23**, 2173.

Twardowski, J. (1978). 'Laser Raman spectroscopy of acid phosphatase from rat liver.' *Biopolymers*, **17**, 181.

Ulmer, K. M. (1983). 'Protein engineering.' *Science*, **219**, 666.

Urabe, H. & Tominaga, Y. (1982). 'Low-lying collective modes of DNA double helix by Raman spectroscopy.' *Biopolymers*, **21**, 2477.

Valleau, J. P. & Torrie, G. M. (1977). 'A guide to Monte Carlo for statistical mechanics: 2. Byways.' In *Statistical Mechanics, Part A: Equilibrium Techniques*, ed. B. J. Berne, pp. 169–94. New York: Plenum.

Valleau, J. P. & Whittington, S. G. (1977). A guide to Monte Carlo for statistical mechanics: 1. Highways.' In *Statistical Mechanics, Part A: Equilibrium Techniques*, ed. B. J. Berne, pp. 137–68. New York: Plenum.

van der Zwan, G. & Hynes, J. T. (1982). 'Dynamical polar solvent effects on solution reactions: A simple continuum model.' *Journal of Chemical Physics*, **76**, 2993.

van Gunsteren, W. F. & Berendsen, H. J. C. (1977). 'Algorithms for macromolecular dynamics and constraint dynamics.' *Molecular Physics*, **34**, 1311.

Vologodskii, A. V., Anshelevich, V. V., Lukashin, A. V. & Frank-Kamenetskii, M. D. (1979). 'Statistical mechanics of supercoils and the torsional stiffness of the DNA double helix.' *Nature*, **280**, 294.

van Gunsteren, W. F. & Berendsen, H. J. C. (1982). 'Molecular dynamics: Perspective for complex systems.' *Biochemical Society Transactions*, **10**, 301.

van Gunsteren, W. F. & Berendsen, H. J. C. (1984). 'Computer simulation as a tool for tracing the conformational differences between proteins in solution and in the crystalline state.' *Journal of Molecular Biology*, **176**, 559.

van Gunsteren, W. F. & Berendsen, H. J. C. (1985). 'Molecular dynamics simulations: Techniques and applications to proteins.' In *Molecular Dynamics and Protein Structure*, ed. J. Hermans, pp. 5–14. Chapel Hill: University of North Carolina.

van Gunsteren, W. F., Berendsen, H. J. C., Geurtsen, R. G. & Zwinderman, H. R. J. (1986). 'A molecular dynamics computer simulation of an eight basepair DNA fragment in aqueous solution: Comparison with experimental 2D NMR data.' *Annals of the New York Academy of Sciences*, in press.

van Gunsteren, W. F., Berendsen, H. J. C., Hermans, J., Hol, W. G. J. & Postma, J. P. M. (1983). 'Computer simulation of the dynamics of hydrated protein crystals and its comparison with X-ray data.' *Proceedings of the National Academy of Sciences of the USA*, **80**, 4315.

van Gunsteren, W. F. & Karplus, M. (1982*a*). Effect of constraints on the dynamics of macromolecules.' *Macromolecules*, **15**, 1528.

van Gunsteren, W. F. & Karplus, M. (1982*b*). 'Protein dynamics in solution and in a crystalline environment: A molecular dynamics study.' *Biochemistry*, **21**, 2259.

Verlet, L. (1967). 'Computer "experiments" on classical fluids. I. Thermodynamical properties of Lennard–Jones molecules.' *Physical Review*, **159**, 98.

Wagner, G. (1983). 'Characterization of the distribution of internal motion in the basic pancreatic trypsin inhibitor using a large number of internal NMR probes.' *Quarterly Reviews of Biophysics*, **16**, 1.

Wagner, G., DeMarco, A. & Wüthrich, K. (1976). 'Dynamics of the aromatic amino acid residues in the globular conformation of the basic pancreatic trypsin inhibitor (BPTI) I. ^{1}H NMR studies.' *Biophysics of Structure and Mechanism*, **2**, 139.

Wagner, G. & Wüthrich, K. (1979). 'Structural interpretation of the amide proton exchange in the basic pancreatic trypsin inhibitor and related proteins.' *Journal of Molecular Biology*, **134**, 75.

Wagner, G. & Wüthrich, K. (1982). 'Amide proton exchange and surface conformation of the basic pancreatic trypsin inhibitor in solution. Studies with two-dimensional nuclear magnetic resonance.' *Journal of Molecular Biology*, **160**, 343.

Wang, A. H. J., Quigley, G. J., Kolpak, F. J., Crawford, J. L., van Boom, J. H., van der Marel, G. & Rich, A. (1979). 'Molecular structure of a left-handed double helical DNA fragment at atomic resolution.' *Nature*, **282**, 680.

Wang, A. H. J., Quigley, G. J., Kolpak, F. J., van der Marel, G., van Boom, J. H. & Rich, A. (1981). 'Left-handed double helical DNA: Variations in the backbone conformation.' *Science*, **211**, 171.

Warshel, A. (1982). 'Dynamics of reactions in polar solvents. Semiclassical trajectory studies of electron-transfer and proton-transfer reactions'. *Journal of Physical Chemistry*, **86**, 2218.

Warshel, A. (1984). 'Dynamics of enzymatic reactions.' *Proceedings of the National Academy of Sciences of the USA*, **81**, 444.

Warshel, A. & Levitt, M. (1976). 'Theoretical studies of enzymic reactions: Dielectric, electrostatic and steric stabilization of the carbonium ion in the reaction of lysozyme.' *Journal of Molecular Biology*, **103**, 227.

Warshel, A. & Russell, S. T. (1984). 'Calculations of electrostatic interactions in biological systems and in solutions.' *Quarterly Review of Biophysics*, **17**, 283.

Warwicker, J. & Watson, H. C. (1982). 'Calculation of the electric potential in the active site cleft due to α-helix dipoles.' *Journal of Molecular Biology*, **157**, 671.

Watson, J. D. & Crick, F. H. C. (1953). 'Molecular structure of nucleic acids.' *Nature*, **171**, 737.

Weber, I. T. & Steitz, T. A. (1984). 'Model of specific complex between catabolite gene activator protein and B-DNA suggested by electrostatic complementarity.' *Proceedings of the National Academy of Sciences of the USA*, **81**, 3973.

Weber, P. C. & Tollin, G. (1985). 'Electrostatic interactions during electron transfer reactions between c-type cytochromes and flavodoxin.' *Journal of Biological Chemistry*, **260**, 5568.

Wegener, W. A. (1982). 'Bead models of segmentally flexible macromolecules.' *Journal of Chemical Physics*, **76**, 6425.

Wegener, W. A., Dowben, R. M. & Koester, V. J. (1980). 'Diffusion coefficients for segmentally flexible macromolecules: General formalism and application to rotational behavior of a body with two segments.' *Journal of Chemical Physics*, **73**, 4086.

Weiner, P. K. & Kollman, P. A. (1981). 'AMBER: Assisted model building with energy refinement. A general program for modeling molecules and their interactions.' *Journal of Computational Chemistry*, **2**, 287.

Weiner, S. J., Kollman, P. A., Case, D. A., Singh, U. C., Ghio, C., Alagona, G., Profeta, S., & Weiner, P. (1984). 'A new force field for molecular mechanical simulation of nucleic acids and proteins.' *Journal of the American Chemical Society*, **106**, 765.

Weiner, S. J., Kollman, P. A., Nguyen, D. T. & Case, D. A. (1986*a*). 'An all atom force field for simulations of proteins and nucleic acids.' *Journal of Computational Chemistry*, **7**, 230.

Weiner, S. J., Seibel, G. L. & Kollman, P. A. (1986*b*). 'The nature of enzyme catalysis in trypsin.' *Proceedings of the National Academy of Sciences of the USA*, **83**, 649.

Weiner, S. J., Singh, U. C. & Kollman, P. A. (1985). 'Simulation of formamide hydrolysis by hydroxide ion in the gas phase and in aqueous solution.' *Journal of the Amerian Chemical Society*, **107**, 2219.

Westhof, E., Altschuh, D., Moras, D., Bloomer, A. C., Mondragon, A., Klug, A. & van Regenmortel, M. H. V. (1984). 'Correlation between segmental mobility and the location of antigenic determinants in proteins.' *Nature*, **311**, 123.

Westhof, E., Dumas, P. & Moras, D. (1985). 'Crystallographic refinement of yeast aspartic acid transfer RNA.' *Journal of Molecular Biology*, **184**, 119.

Westhof, E., Gallion, S. L., Weiner, P. K. & Levy, R. M. (1986). 'Temperature dependent molecular dynamics and restrained X-ray refinement simulations of a Z-DNA hexamer.' *Journal of Molecular Biology*, in press.

Westhof, E. & Sundaralingam, M. (1980). 'Interrelationships between the pseudorotation parameters P and τ_m and the geometry of the furanose ring.' *Journal of the American Chemical Society*, **102**, 1493.

Westhof, E. & Sundaralingam, M. (1983*a*). 'Disorder and dynamics of five-membered rings in biological structures.' *Structure and Dynamics: Nucleic Acids and Proteins*, eds E. Clementi & R. H. Sarma, pp. 135–47. New York: Adenine Press.

Westhof, E. & Sundaralingam, M. (1983*b*). 'A method for the analysis of puckering disorder in five-membered rings: The relative mobilities of furanose and proline rings and their effects on polynucleotide and polypeptide backbone flexibility.' *Journal of the American Chemical Society*, **105**, 970.

Whitesides, G. M. & Wong, C. H. (1985). 'Enzymes as catalysts in synthetic organic chemistry.' *Angewandte Chemie, International Edition in English*, **24**, 617.

Wickstrom, E., Behlen, L. S., Reuben, M. A. & Ainpour, P. R. (1981).

'Molecular rulers for measuring RNA structure: Sites of crosslinking in chlorambucilyl-phenylalanyl-tRNA^{Phe} (yeast) and chlorambucilyl-pentadecaprolyl-phenylalanyl-tRNA^{Phe} (yeast) intramolecularly crosslinked in aqueous solution.' *Proceedings of the National Academy of Sciences of the USA*, **78**, 2082.

Widom, J. (1985). 'Bent DNA for gene regulation and DNA packaging.' *BioEssays*, **2**, 11.

Williams, R. J. P. (1980). 'On first looking into nature's chemistry. Part II. The role of large molecules, especially proteins.' *Chemical Society Reviews*, **9**, 325.

Willis, B. T. M. & Pryor, A. W. (1975). *Thermal Vibrations in Crystallography*. Cambridge University Press.

Wilson, E. B., Decius, J. C. & Cross, P. C. (1955). *Molecular Vibrations*. New York: McGraw-Hill.

Wing, R., Drew, H., Takano, T., Broka, C., Tanaka, S., Itakura, K. & Dickerson, R. E. (1980). 'Crystal structure analysis of a complete turn of B-DNA.' *Nature*, **287**, 755.

Wlodawer, A. & Sjölin, L. (1982). 'Hydrogen exchange in RNase A: Neutron diffraction study.' *Proceedings of the National Academy of Sciences of the USA*, **79**, 1418.

Wolfenden, R. (1983). 'Waterlogged molecules.' *Science*, **222**, 1087.

Wolynes, P. G. (1980). 'Dynamics of electrolyte solutions.' *Annual Review of Physical Chemistry*, **31**, 345.

Wolynes, P. G. (1984). 'Chemical physics of molecular systems in condensed phases.' In *Monte Carlo Methods in Quantum Problems*, ed. M. H. Kalos, pp. 71–104. Holland: D. Reidel.

Wolynes, P. G. & McCammon, J. A. (1977). 'Hydrodynamic effect on the coagulation of porous biopolymers.' *Macromolecules*, **10**, 86.

Wong, C. F. & McCammon, J. A. (1986*a*). 'Dynamics and design of enzymes and inhibitors.' *Journal of the American Chemical Society*, **108**, 3830.

Wong, C. F. & McCammon, J. A. (1986*b*). 'Computer simulation and the design of new biological molecules.' *Israel Journal of Chemistry*, in press.

Woo, N. H., Roe, B. A. & Rich, A. (1980). 'Three-dimensional structure of Escherichia coli initiator $tRNA_f^{Met}$.' *Nature*, **286**, 346.

Wood, D. W. (1979). 'Computer simulation of water and aqueous solutions.' In *Water: A Comprehensive Treatise*, ed. F. Franks, p. 279. New York: Plenum.

Wood, W. W. (1968). In *Physics of Simple Liquids*, eds H. N. V. Temperley, G. S. Rushbrooke & J. S. Rowlinson, p. 116. Amsterdam: North-Holland.

Woodward, C. K. & Hilton, B. D. (1979). 'Hydrogen exchange kinetics and internal motions in proteins and nucleic acids.' *Annual Review of Biophysics and Bioengineering*, **8**, 99.

Wright, H. T., Manor, P. C., Beurling, K., Karpel, R. L. & Fresco, J. R. (1979). 'The structure of baker's yeast tRNA: A second tRNA conformation.' In *Transfer RNA: Structure, Properties and Recognition*, eds. P. R. Schimmel, D. Söll & J. N. Abelson, pp. 145–60. New York: Cold Spring Harbor.

Wüthrich, K. & Wagner, G. (1979). 'Nuclear magnetic resonance of labile protons in the basic pancreatic trypsin inhibitor.' *Journal of Molecular Biology*, **130**, 1.

Yguerabide, J., Epstein, H. F. & Stryer, L. (1970). 'Segmental flexibility in an antibody molecule.' *Journal of Molecular Biology*, **51**, 573.

Young, R. D. & Bowne, S. F. (1984). 'Conformational substates and barrier height distributions in ligand binding to heme proteins.' *Journal of Chemical Physics*, **81**, 3730.

Yu, H. A., Karplus, M. & Hendrickson, W. A. (1985). 'Restraints in temperature-factor refinement for macromolecules: An evaluation by molecular dynamics.' *Acta Crystallographica*, **B41**, 191.

Zauhar, R. J. & Morgan, R. S. (1985). 'A new method of computing the macromolecular electric potential.' *Journal of Molecular Biology*, **186**, 815.

Zwanzig, R. & Ailawadi, N. K. (1969). 'Statistical error due to finite time averaging in computer experiments.' *Physical Review*, **182**, 280.

INDEX

LANCASTER UNIVERSITY LIBRARY

Due for return by end of service on date below
(or earlier if recalled)

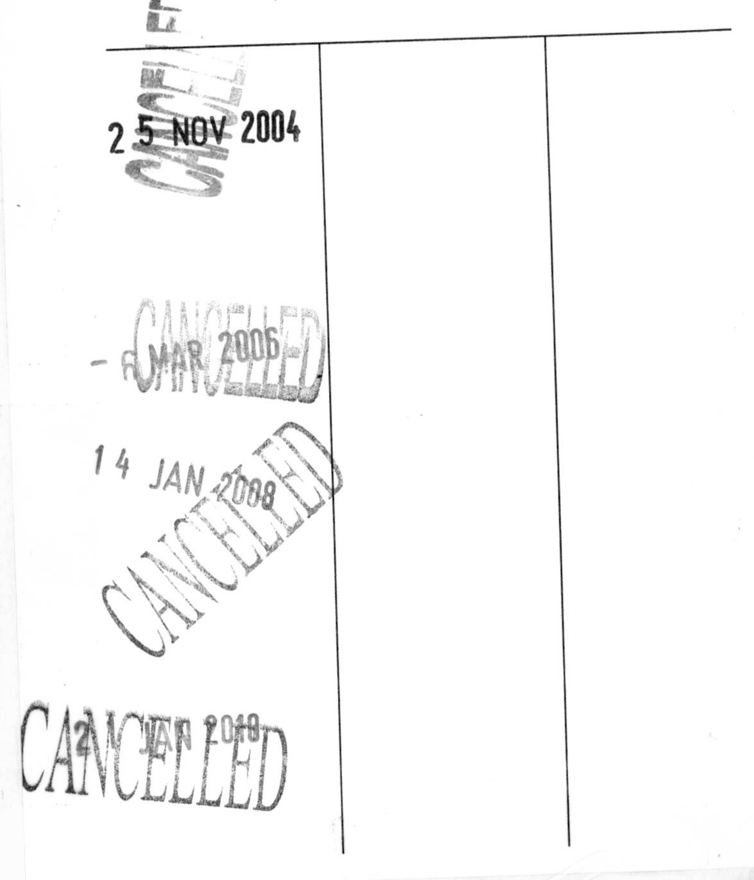